Praise for

Solar Electricity Basics

Dan effectively takes the guess-work out of solar energy with simple explanations and instructions that anyone can follow. Just as importantly he presents all the options in a way that makes deciding which installation might best suit your needs. This is a must-read for anyone looking to make steps towards a resilient energy future.

— Oliver M. Goshey, founder and director, Abundant Edge

Lucky for you, Dan Chiras strikes again. With his usual plain-spoken clarity, Dan makes a complex subject easy and accessible. The perfect guide for anyone to go solar!

— Bruce King, author, *The New Carbon Architecture*

Thinking of going solar? Even if you intend to hire experts to install your system, it helps enormously to know a little about the technology, and how to assess your site's solar resource. Dan Chiras is the perfect guide; he's enthusiastic, knowledgeable, and makes it all easy to understand.

— Richard Heinberg, Senior Fellow, Post Carbon Institute, and author, *The End of Growth*

Dan Chiras is the man who launched a thousand rooftop solar arrays. Don't even think about dabbling in photovoltaics without Dan's guidance. *Solar Electricity Basics, Second Edition* is the essential go to manual for getting a grip on the ins and outs of photovoltaic systems. Read it, use it, and harness the power of the sun on your home, farm, or homestead.

— John D. Ivanko and Lisa Kivirist, co-authors, *ECOpreneuring*

SOLAR ELECTRICITY Basics

2ND REVISED EDITION

SOLAR ELECTRICITY Basics

POWERING YOUR HOME OR OFFICE WITH SOLAR ENERGY

DAN CHIRAS

Cover design by Diane McIntosh.
Cover Images © iStock.

Printed in Canada. First printing November, 2019.

Inquiries regarding requests to reprint all or part of *Solar Electricity Basics, Revised & Updated 2nd Edition* should be addressed to New Society Publishers at the address below. To order directly from the publishers, please call toll-free (North America) 1-800-567-6772, or order online at www.newsociety.com

Any other inquiries can be directed by mail to:
New Society Publishers
P.O. Box 189, Gabriola Island, BC V0R 1X0, Canada
(250) 247-9737

LIBRARY AND ARCHIVES CANADA CATALOGUING IN PUBLICATION

Title: Solar electricity basics : powering your home or office with solar energy / Dan Chiras.

Names: Chiras, Daniel D., author.
Description: 2nd revised edition. | Includes index.
Identifiers: Canadiana (print) 20190145528 | Canadiana (ebook) 20190145536 | ISBN 9780865719255
(softcover) | ISBN 9781550927184 (PDF) | ISBN 9781771423144 (EPUB)
Subjects: LCSH: Photovoltaic power generation—Handbooks, manuals, etc. | LCSH: Solar energy—
Handbooks, manuals, etc. | LCGFT: Handbooks and manuals.
Classification: LCC TK10857 .C56 2019 | DDC 621.31/244—dc23

Funded by the Government of Canada | Financé par le gouvernement du Canada | Canada

New Society Publishers' mission is to publish books that contribute in fundamental ways to building an ecologically sustainable and just society, and to do so with the least possible impact on the environment, in a manner that models this vision.

Contents

CHAPTER 1

An Introduction to Solar Electricity

The world is in the midst of a global renewable energy revolution. With installations of new solar and wind energy systems climbing rapidly, nearly 50 countries have pledged to be 100% dependent on renewable energy by 2050. That's no pipe dream. It's highly doable. In fact, researchers at Stanford University project that the world could be 100% reliant on renewables in 20 to 40 years. The technologies and know-how are here, and in many parts of the world, so is the commitment.

Although the United States and Canada are not leading the race to a renewable energy future, significant progress is being made in both countries. In the US, for instance, solar, wind, biomass, hydroelectricity, and geothermal now produce about 17% of the nation's electricity—up from 7 to 8 percent at the beginning of the millennium.

One reason why renewable energy production has increased in recent years in the US is declining cost. Because of this, solar electric systems now consistently generate electricity at or below the cost of power from conventional sources like nuclear power and coal. With federal tax credits that have been in place for over a decade, solar electric systems on homes and businesses consistently produce electricity much cheaper than utilities. Declining costs and rising popularity have led many utilities to install solar and wind farms.

In this book, we will explore solar electric systems, technically known as *photovoltaic systems* or *PV systems,* for short. I will primarily present information on residential-scale PV systems—systems suitable for homes and small businesses. These systems generally fall in the range of 1,000 watts for very small, energy-efficient cabins or cottages to 5,000 to 15,000 watts for typical suburban homes. All-electric homes could require even larger systems (on the order of 25,000 watts). Small businesses typically require much larger systems. Before we move on, though, what does it mean when I (or a solar installer) refer to a 10,000-watt system?

Sizing Solar Systems: Understanding Rated Power

Solar electric systems are sized by their wattage. As just noted, a home might require a 10,000-watt system. The wattage rating is known as the *rated power* of a system. Rated power is also known as *rated capacity.*

The rated power of a system is the rated power of the modules (frequently, but improperly, referred to as *panels*) multiplied by the number of modules in the system. If your system contains ten 300-watt solar modules, its rated power, or capacity, is 3,000 watts. A system with twenty 300-watt modules is rated at 6,000 watts. To simplify matters, installers and other professionals convert watts to kilowatts. So, a 10,000-watt system is a 10-kilowatt (or kW) system. That's a shorthand measure based on the fact that kilo in scientific jargon means 1,000.

But what does it mean when I say a *module's* rated power is 300 watts?

The rated power of a solar module is the instantaneous output of the module measured in watts under an industry-established set of conditions known as *standard test conditions (STC)*. By carefully controlling the temperature of the modules and the intensity of the light they are exposed to, all manufacturers rate their modules similarly. That way, buyers like you know what they're getting—as do installers and system designers. But what does it really mean that a module is rated at 300 watts?

Of course, most of us are familiar with the term *watts.* When shopping for light bulbs, for instance, we select them based on wattage. The higher the wattage, the brighter the bulb. We also compare microwave ovens and hair dryers by their wattage. The higher the wattage, the more energy these devices consume. In these and other electronic devices, wattage is a measure of power consumption.

Technically speaking, though, wattage is the rate of the flow of energy. The higher the wattage, the greater the flow. So, a 100-watt incandescent light bulb will use a lot more power each second it operates than a 12-watt LED light.

Watts is also used to rate technologies that *produce* electricity, such as solar modules or power plants. For example, most solar modules installed these days are rated at 275 to 300 watts. Power plants are rated in millions of watts, or megawatts; a typical coal-fired power plant produces 500 to 1,000 megawatts of power.

To test a solar module to determine its rated power, workers set it up in a room that is maintained at 77°F (25°C). A light is flashed on the module at an intensity of 1,000 watts per square meter. That's equivalent to full sun on a cloudless day in most parts of the world. The module is arranged so that light rays strike it at a 90° angle, that is, perpendicularly. (Light rays striking the module perpendicular to its surface result in the greatest absorption of the light.) A 300-watt module will produce 300 watts of energy under such conditions.

To size a solar electric system, solar installers first determine the size of the system (in watts) a customer needs to meet his or her needs. (I'll explain that process later.) They then divide that number—the rated power of the system—by the rated power of the modules they're going to use. This simple math reveals the number of modules they will need to install. If a customer needs a 9,000-watt system, for instance, he or she would require thirty 300-watt modules.

An Overview of Solar Electric Systems

Now that you understand a little bit about solar system sizing, let's take a look at solar electric systems.

Solar electric systems capture sunlight energy and convert it into electrical energy. This conversion takes place in the solar modules, more commonly referred to as *solar panels.* A solar module consists of numerous *solar cells.* Most solar modules contain 60 or 72 solar cells.

Solar cells are made from one of the most abundant chemical substances on Earth: silicon dioxide. It makes up 26% of the Earth's crust. Silicon dioxide is found in quartz and a type of sand that contains quartz particles. Silicon for module construction is extracted from sand.

Solar cells are wired to one another in a module. The modules are encased in glass (on the front) and usually a thin layer of plastic sheeting (on the back) to protect the cells from the elements, especially moisture. Most modules have silver or black aluminum frames, although there are some frameless modules on the market, which were introduced for aesthetics and to simplify installation and reduce cost.

Two or more modules are typically mounted on a rack and wired together. Together, the rack and modules are known as a *solar array.*

Figure 1.1. *These solar modules being installed by several students at The Evergreen Institute consist of numerous square solar cells. Each solar cell has a voltage of around 0.6 volts. The cells are wired in series to produce higher voltage, which helps move the electrons from cell to cell. Modules are also wired in series to increase voltage.* Credit: Dan Chiras.

Ground-mounted arrays are usually anchored to the earth by a steel-reinforced concrete foundation. In such cases, the foundation is also considered part of the array.

How Solar Cells and Modules Are Wired

Inside a solar module, solar cells are "wired" together in series. The modules in an array are then also wired in series. What does all this mean?

Series wiring is a technique commonly used in electrical devices. Batteries in a flashlight, for instance, are placed so that the positive end of one battery contacts the negative end of the next one. That's series wiring. It's done to increase the voltage. When you place two 1.5-volt batteries in series in a flashlight, you increase the voltage to 3 volts. Place four 1.5-volt batteries in series, and the voltage is 6.

So, what's voltage?

Voltage is the somewhat mysterious force that moves electrons in wires and electrical devices. The higher the voltage, the greater the force.

In solar modules, each solar cell has a positive and a negative lead (wire). When manufacturing solar cells, the positive lead of one solar cell is soldered to the negative of the next, and so on and so on.

Most solar cells in use today are square and measure 5 × 5 inches (125 × 125 mm) or 6 × 6 inches (156 × 156 mm) and have a voltage of about 0.6 volts. When manufacturers wire (actually, solder) 60 solar cells together in series in a module, the voltage increases to 36. Seventy-two cells wired in series creates a 43-volt solar module.

Solar modules are themselves wired in series to further increase voltage. Ten 36-volt modules wired in series results in an array voltage of 360. Two more modules added to the series brings the voltage to 432 volts. Higher voltages allow us to transmit electricity more efficiently over greater distances.

A group of modules wired in series is known as a *series string.* Most residential solar arrays consist of one or more series strings, each containing 10 to 12 modules. String sizes in commercial PV arrays are larger.

Converting DC Electricity into AC Electricity

Solar modules produce direct current (DC) electricity. That's the kind of electricity released by all batteries.

DC electricity is one of two types of electricity in use today. It consists of a flow of tiny subatomic particles called *electrons* through conductors, usually copper wires. In DC electricity, electrons flow in one direction (Figure 1.2). As illustrated, energized electrons stored in a charged battery flow out of the battery along a wire through a metal filament in a light bulb. Here, the energy of those electrons is used to create light and heat. The de-energized electrons then flow back into the battery.

In a solar system, DC electricity produced in the solar modules flows (via wires) to a device known as an *inverter.* Its job is to convert DC electricity produced by solar cells into alternating current (AC) electricity, the type of electricity used in homes and businesses throughout most of the world. In wires and devices that are powered by AC electricity, electrons flow back and forth—that is, they *alternate* direction very rapidly.

Figure 1.2. *Solar electric systems produce direct current (DC) electricity just like batteries. As shown here, in DC circuits the electrons flow in one direction. The energy the electrons carry is used to power loads like light bulbs, heaters, and electronics.* Credit: Forrest Chiras.

You may have noticed that the solar electric system I've just described has no batteries. Many people erroneously think that all solar systems include batteries. While that was true in earlier times, battery-based solar electric systems are extremely rare these days; they aren't much needed anymore. The vast majority of systems in more developed countries are connected to the electrical grid—the wires that feed electricity to our communities. When a solar electric system is generating electricity, it powers the home or business. Any surplus electricity it produces is fed onto the grid, where it is consumed by one's neighbors. However, the utility keeps track of that surplus and gives it back to the producer—you or me—free of charge whenever we need it. Although it's a bit more complicated than this, you can think of the grid as a very large battery. (More on this later.)

Applications

Once a curiosity, solar electric systems are becoming commonplace throughout the world. Besides being installed on homes, they are now powering schools, small businesses, factories, and office buildings. Microsoft, Toyota, and Google are powered by large solar electric systems. More and more electric utilities are installing large PV arrays. In fact, utility-scale solar systems make up a large portion of the industry today.

Even colleges are getting in on the act. Colorado College (where I once taught courses on renewable energy) installed a large solar system on one of its dormitories. A local middle school hired me to install an array on its grounds (Figure 1.3). Several airports, notably those in Sacramento,

Figure 1.3. *This solar array was installed at the Owensville Middle School in Owensville, Missouri, by the author and his business partner, Tom Bruns, and numerous hard-working students. The array is small for a school of this size, which uses almost 1,000 kilowatt-hours per day; the array only provides about 12 days' worth of electricity to the school. The array is also used for science and math education.* Credit: Dan Chiras.

Denver, and Indianapolis, are partially powered by large solar arrays, making use of the vast amounts of open space that surround them.

Farmers and ranchers install solar electric systems to power electric fences or to pump water for their livestock, saving huge amounts of money on installation. My wife and I power our farm and our home entirely with solar and wind energy. We even installed a solar pond aerator to reduce sludge buildup, lengthen the lifespan of our pond, and keep it open during the winter (Figure 1.4).

Figure 1.4. (a) *Aerating my half-acre farm pond reduces bottom sludge and helps create a healthier environment for the fish.* (b) *To prevent the pond from icing over completely in the winter (so our ducks can swim), I designed and installed a solar pond aerator in conjunction with a St. Louis company, Outdoor Water Solutions. They are currently selling a design like this online.*
Credit: Dan Chiras.

Solar electric systems are used on sailboats and other watercraft to power lights, fans, radio communications, GPS systems, and refrigerators. Many recreational vehicles (RVs) are equipped with small PV systems as well.

Bus stops, parking lots, and highways are now often illuminated by superefficient lights powered by solar electricity, as are many portable information signs used at highway construction sites (Figure 1.5). Numerous police departments now set up solar-powered radar units in neighborhoods. These units display the speed of cars passing by and warn drivers if they're exceeding the speed limit.

You'll find solar-powered trash compactors at the Kennedy Space Center and in at least 1,500 cities in the US. In downtown Colorado Springs, sidewalk parking meter pay stations are similarly powered. In Denver, you'll find solar-powered credit card payment centers for city bikes that people can rent for the day (Figure 1.6).

Backpackers and river runners can take small solar chargers or larger lightweight fold-up solar modules with them on their ventures into the

Figure 1.5. *This electronic road sign is powered by a small solar array.*
Credit: Dan Chiras.

Figure 1.6. *These bikes in downtown Denver, Colorado, can be rented by the hour. The payment system (left) is powered by solar electricity.* Credit: Dan Chiras.

wild to power small electronic devices like cell phones. Military personnel have access to similar products, and there are numerous portable devices available for charging cell phones and tablets in our daily lives. Some are even sewn into backpacks, sports bags, or briefcases. There are also a number of cart and trailer-mounted PV systems on the market that can supply emergency power after disasters.

Solar electric systems are well suited for remote cottages, cabins, and homes where it is often cost prohibitive to run power lines. In France, the government paid to install solar electric systems and wind turbines on farms at the base of the Pyrenees Mountains rather than running electric lines to these distant operations.

PV systems are becoming very popular in less-developed countries. They are, for instance, being installed in remote villages to power lights and computers as well as refrigerators and freezers that store vaccines and other medicine. They're also used to power water pumps.

Solar electricity is also used to power remote monitoring stations that collect data on rainfall, stream flow, temperature, and seismic activity, and PVs allow scientists to transmit data back to their labs from such sites.

Solar electric modules often power lights on buoys, vital for nighttime navigation on large rivers like the Saint Lawrence River. Railroad signals and aircraft warning beacons are also often solar-powered.

The ultimate in remote and mobile applications, however, has to be the satellite. Virtually all military and telecommunications satellites are powered by solar electricity, as is the International Space Station.

World Solar Energy Resources

Solar electricity is rapidly growing in popularity, which is fortunate for all of us because global supplies of the fossil fuels we have long used to generate electricity such as coal, natural gas, and oil are on the decline. As supplies wane, and as efforts to address global climate change increase, solar electric systems—along with small and large wind systems and other forms of renewable energy farms—could very well become a major source of electricity throughout the world. Frankly, I think it is inevitable. But is there enough solar energy to meet our needs?

Although solar energy is unevenly distributed over the Earth's surface, significant resources are found on every continent. "Solar energy's potential is off the chart," write energy experts Ken Zweibel, James Mason, and Vasilis Fthenakis in a December 2007 article, "A Solar Grand Plan," published in *Scientific American.* Less than one billionth of the Sun's energy strikes the Earth, but, as they point out, the solar energy striking the Earth in a 40-minute period is equal to all the energy human society consumes in a year. That is, 40 minutes of solar energy is equivalent to all the coal, oil, natural gas, oil shale, tars sands, hydropower, and wood we consume in an entire year. What is more, we'd only need to capture about 0.01% of the solar energy striking the Earth to meet *all* of our energy demands. Clearly, solar electric systems mounted on our homes and businesses or in giant commercial solar arrays could tap into the Sun's generous supply of energy, providing us with an abundance of electricity.

Most renewable energy experts envision a system that consists of dramatic improvements in efficiency and a mix of renewable energy technologies. Large-scale wind farms will very likely provide a large percentage of the world's electricity. (The world currently produces three times as much electricity from wind as it does from solar.) Geothermal and biomass resources will contribute to the world's energy supply. *Biomass* is plant matter that can be burned to produce heat or to generate steam that's used to power turbines that generate electricity. Biomass can also be converted to liquid or gaseous fuels that can be burned to produce electricity or heat or power vehicles. Hydropower currently contributes a significant amount of electricity throughout the world, and it will continue to add to the energy mix in our future.

What will happen to conventional fuels such as oil, natural gas, coal, and nuclear energy? Although their role will diminish over time, these fuels will probably contribute to the energy mix for many years to come. In the future, however, they will very likely take a back seat to renewables. They could eventually become pinch hitters to renewable energy, supplying backup electricity to an otherwise renewable energy-powered system.

Despite what some ill-informed critics say, renewable energy is splendidly abundant. What is more, the technologies needed to efficiently capture and convert solar energy to useful forms of energy like heat, light, and electricity are available now and, for the most part, they are quite affordable.

And don't lose track of this fact: While nuclear and fossil fuels are on the decline, the Sun will enjoy a long future. The Sun, say scientists, will continue to shine for at least five billion more years. The full truth, however, is that scientists predict the Sun's output will increase by about 10% in one billion years, making planet Earth too hot to sustain life. So, we won't be enjoying five billion more years of sunshine; we have less than a billion years before we'll have to check out. But that's a bit more than the 30 to 50 years of oil we have left.

What the Critics Say

Proponents of a solar-powered future view solar energy as an ideal fuel source. It's clean, and it's relatively inexpensive because the fuel is free. Solar is also abundant and will be available for a long, long time. Moreover, its use could ease many of the world's most pressing problems, such as global climate change.

Like all fuels, solar energy is not perfect. Critics like to point out that, unlike conventional resources such as coal, the Sun is not available 24 hours a day. Some people don't like the looks of solar electric systems. And, many uninformedly argue that solar is too expensive. Let's take a look at the most common arguments and respond to the criticisms.

Availability and Variability

Although the Sun shines 24 hours a day and beams down on the Earth at all times, half the planet is always blanketed in darkness. This poses a problem for humankind, especially those of us in the more developed countries, as we consume electricity 24 hours a day, 365 days a year.

Another problem is the daily variability of solar energy. That is, even during daylight hours, clouds can block the sun, sometimes for days on end. At night, PVs produce no energy at all. If solar electric systems are unable to generate electricity 24 hours a day like coal-fired and natural-gas-fired power plants, how can we use them to meet our 24-hour-per-day demand for electricity?

Some solar homeowners who live off grid (that is, not connected to the electrical grid) solve the problem by using batteries to store the electricity needed to meet nighttime and cloudy-day demand. I lived off grid for 14 years, and had electricity 24 hours a day, 365 days a year—almost entirely supplied by my PV system. I met nighttime and cloudy-day demand thanks to my batteries. I rarely ran out of electricity, and when I did, I fired up my backup generator.

Although efforts are being made to build large storage facilities, I doubt that batteries could serve as a backup for modern society. We use—and waste—way too much electricity. But don't dismay. Solar's less-than-24-hour-per-day availability can also be offset by coupling solar electric systems with other renewable energy sources, for example, wind-electric systems or hydroelectric systems.

Wind systems in good locations often generate electricity day and night (Figure 1.7). It can be transported by the national electrical grid to regions hundreds of miles away, making up for times when solar energy is reduced by clouds or eliminated by nightfall. Figure 1.8 shows how Denver, Colorado, can be supplied by wind and solar from neighboring areas.

Hydropower, generating electricity from flowing water, can also be used to provide a steady supply of electricity in a renewable energy future. In Canada, for example, hydroelectric facilities are treated as a *dispatchable energy resource,* much like natural gas is today in the United States. That is, they can be turned on and off as needed to meet demand. In a finely tuned power system, wind, solar, hydropower, and other renewables could be used to provide a steady supply of electricity day and night. Coal and natural gas could provide backup.

Figure 1.7. *The author's boys, Skyler and Forrest, check out a large wind turbine at a wind farm in Canastota, New York. Wind farms like this one are popping up across the nation—indeed, across the world—producing clean, renewable electricity to power our future.* Credit: Dan Chiras.

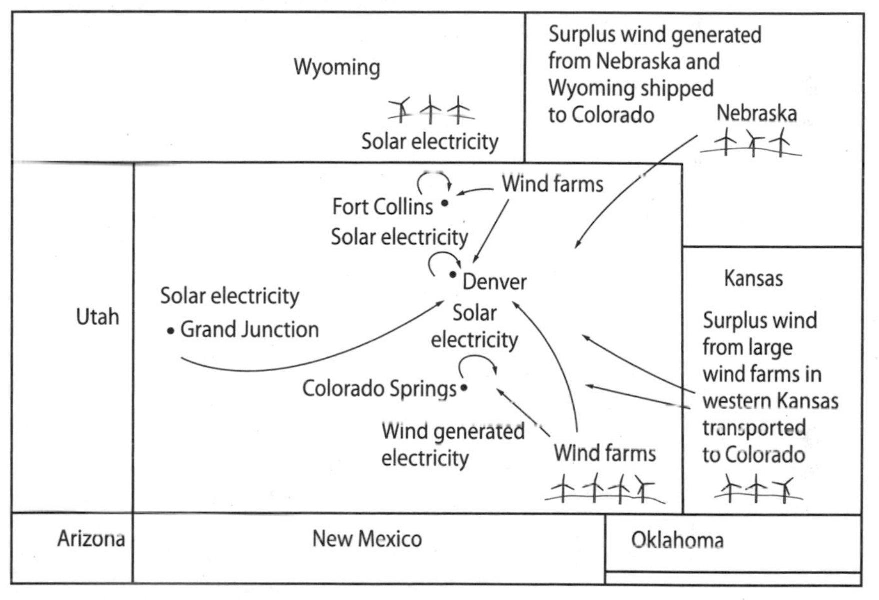

Figure 1.8. *Electricity can be transported from one state or province to another via the electrical grid, a network of high-voltage transmission lines. As illustrated here, surplus electricity from solar and wind energy systems can be imported from neighboring states or from nearby areas, helping to create a steady supply of electricity through rain and shine.* Credit: Forrest Chiras.

With smart planning, forecasting, and careful design, we can meet a good portion of our electrical needs from renewables with the remainder coming from our pinch hitters, the nonrenewables. European nations such as Denmark, Germany, and Spain are already successfully integrating renewable energy into their electrical grids on a large scale.

Aesthetics

While many of us view a solar electric array as a thing of great value, even beauty, some don't. Your neighbors, for instance, might think that PV systems detract from the beauty of their neighborhood.

Ironically, those who object to solar electric systems rarely complain about visual blights like cell phone towers, electric transmission lines, and billboards. One reason that these common eyesores draw little attention is that they have been present in our communities for decades. We've grown used to ubiquitous electric lines and cell phone towers. But PV arrays are relatively new, and people aren't used to them yet.

Fortunately, there are ways to mount a solar array so that it blends with a roof. As you'll learn in Chapter 8, solar modules can be mounted 6 inches (15 cm) or so off, and parallel to, the roof surfaces (Figure 1.9). Manufacturers are also producing much-less-conspicuous modules with black frames or no frames at all, as noted earlier. There's also a solar product called PV laminate that is applied directly to the type of metal roof

Figure 1.9. *Solar arrays can be mounted close to the roof to reduce their visibility. Like the arrangement shown here, arrays can be supported by aluminum rails, usually about 6 inches (15 cm) above the surface of the roof (the gap helps cool them in the summer). While attractive, arrays installed this way tend to produce less electricity than pole-mounted or ground-mounted PV arrays, which stay much cooler on hot summer days.* Credit: Rochester Solar Technologies.

known as *standing seam metal roofing.* This results in an even lower-profile array (Figure 1.10). Solar arrays can sometimes be mounted on poles or racks anchored to the ground that can be placed in sunny backyards—out of a neighbor's line of sight (Figure 1.11).

Figure 1.10. *This product, known as PV laminate, is a plastic-coated flexible material that adheres to standing seam metal roofs. It's best applied to new roof panels before they are installed on the building. PV laminate is not as sensitive to high temperatures as conventional silicon-crystal modules used in most solar systems.* Credit: Uni-Solar.

Figure 1.11. *Ground-mounted solar arrays like this one accommodate numerous modules and can be oriented and angled to maximize production. Because they allow air to circulate freely around the modules, systems such as these stay cooler than many roof-mounted arrays and thus tend to have a higher output.* Credit: Dan Chiras.

Cost

For years, the biggest disadvantage of solar electric systems has been the cost. Those of us who advocated for solar electricity in the 1970s through the 1990s had to appeal to people's sense of right and wrong and their understanding of the external economic costs—that is, the environmental costs—of conventional power. Thankfully, those days are over. As noted earlier, solar electricity is now cost competitive—and frequently cheaper—over the long haul than electricity from conventional sources, even in rural areas. I'll show you the proof of this assertion in Chapter 4.

Solar electricity makes sense in most areas with a decent amount of sunshine. In those areas with abundant sunshine and high electricity costs, like southern California or Hawaii, it makes even more sense. Solar electricity may also make sense for those who are building homes in remote locations a long way from power lines. In such instances, utilities often charge customers gargantuan fees to connect to their power lines. A homeowner, for instance, could pay $20,000 to connect, even when building a home only 0.2 miles from a line.

PV systems produce electricity far cheaper than power companies, but here's the rub: there's a rather large upfront cost. This is what makes solar electricity seem so expensive. When buying a solar electric system, you are paying for the technology that will produce all of the electricity you'll consume over the next 30 to 50 years. That said, this investment often earns a return on investment ranging from 5% to 7% per year.

To avoid the high cost, some individuals lease solar electric systems. Many companies will install solar systems for free, then charge a monthly fee. Individuals can also buy into community solar systems—that is, solar electric systems installed in other locations, like nearby solar farms. Homeowners and business owners purchase solar modules from the farm. The output of the modules a customer pays for at these sites is then credited to his or her electric bill. More on this later.

The Advantages of Solar Electric Systems

Although solar electricity—like any fuel source or technology—has some downsides, they're not insurmountable, and they are outweighed by the many advantages. One of the most important advantages is that solar energy is an abundant, renewable resource—one that will be with us for hundreds of thousands of years. While natural gas, oil, coal, and nuclear fuels are limited and on the decline, solar energy will be available to us for at least one billion years.

Solar energy is a clean resource, too. By reducing the world's reliance on coal-fired power plants, solar electricity could help us reduce our contribution to a host of environmental problems, among them acid rain, global climate disruption, habitat destruction, species extinction, cropland loss caused by desertification, and mercury contamination of surface waters (caused by the release of mercury from coal-fired power plants). Solar electricity could even replace costly nuclear power plants.

Solar energy could help us decrease our reliance on declining and costly supplies of fossil fuels like coal, natural gas, and oil. Although very little electricity in the US comes from oil, electricity generated by solar electric systems could someday be used to power electric vehicles like the Tesla, Chevy Bolt, or Nissan Leaf—and the dozen or so other models being produced by the world's auto manufacturers. In such cases, solar will serve as a clean replacement for oil, the source of gasoline.

Without a doubt, the production of solar electric systems does have its impacts, but it is a relatively benign technology compared to fossil-fuel and nuclear power plants.

Figure 1.12. *Dan with His Electric Car. The author's all-electric 2013 Nissan Leaf, which he powers entirely with solar and wind energy. Folks interested in buying a used all-electric car can get some killer deals on slightly used vehicles. This car costs over $35,000 new. He got it in January 2016 for under $9,000, and it was in perfect condition.* Credit: Linda Stuart-Chiras.

Another benefit of solar electricity is that, unlike oil, coal, natural gas, and nuclear energy, the fuel is free. Moreover, solar energy is not owned or controlled by hostile foreign states or one of the dozen or so major energy companies that strongly influence global energy policy. Because the fuel is free, solar energy can provide a hedge against inflation, which is fueled in part by ever-increasing fuel costs. That's why solar is popular among many businesses and some school districts. Put another way, solar is an investment for a lifetime of inexpensive electricity not subject to annual price hikes.

An increasing reliance on solar and wind energy could also ease political tensions worldwide. Solar and other renewable energy resources could alleviate the perceived need for costly military operations aimed at stabilizing (controlling the politics of) the Mideast, a region where the largest oil reserves reside. Not a drop of human blood will be shed to ensure the steady supply of solar energy to fuel our economy.

Yet another advantage of solar-generated electricity is that it uses existing infrastructure (the electrical grid). A transition to solar electricity and wind energy could occur fairly seamlessly.

Solar electricity is also modular. (That's one reason why solar "panels" are called "modules.") That is, you can build a system over time. If you can initially only afford a small system, you can start small and expand your system as money becomes available. Expandability has been made easier by the invention of *microinverters,* small inverters that are wired to each solar module in a PV system. (For years, most solar systems were designed with one inverter, known as a *string inverter.* If you wanted to add modules, you'd almost always have to buy a larger, more expensive inverter. With microinverters, a homeowner can add one module and one microinverter at a time. It is much more economical to expand this way.)

Solar electricity does not require extensive use of water, as do coal, nuclear, and gas-fired power plants.

On a personal level, solar electric systems offer considerable economic savings over their lifetime, a topic discussed in Chapter 4. They also create a sense of pride and accomplishment, and they generate tremendous personal satisfaction.

A 2015 report by the Lawrence Berkeley National Laboratory on home sales in eight states from 2002 to 2013 showed that solar electric systems boosted selling prices. Sales prices on homes with PV systems were about $4.17 per watt higher than comparable "solarless" homes. If you had a 10-kW PV array on your home, you would have made an additional $41,700. That's amazing, especially when you take into account that the cost of a PV system nationwide at the time of the study was about $3.46 per watt, installed. That system would have cost $34,600 up front. Had you availed yourself to the 30% federal tax credit, the cost would have been $24,200. So not only would you have almost doubled your money, the system would have generated a ton of electricity free of charge.

Purpose of This Book

This book focuses primarily on solar electric systems for homes and small businesses. It is written for individuals who aren't well versed in electricity. I'll teach you what you need to know about electricity as we go along, so you won't need a master's degree in electrical engineering to understand the material on solar systems.

When you finish reading and studying the material in this book, you will have the knowledge required to assess your electrical consumption and the solar resource at your site. You will also be able to determine if a solar electric system will meet your needs and if it makes sense for you. You will know what kind of system you should install, and you'll have a good working knowledge of the key components of PV systems. You'll also know how PV systems are installed and what their maintenance requirements are.

This book will help you know what to look for when shopping for a PV system to install yourself or when talking to a professional installer. If you choose to hire a professional to install a system, you'll be thankful you've read and studied the material in this book. The more you know, the more informed input you can provide into your system design, components, siting, and installation—and the more likely you'll be happy with your purchase.

I should point out, however, that this book is not an installation manual. Reading it won't qualify you to install a solar electric system. It is a good start, however.

Organization of This Book

We'll begin our exploration of solar electricity in the next chapter by studying the Sun and solar energy. I will discuss important terms and concepts such as average peak sun hours. You'll learn why solar energy varies during the year, and how to calculate the proper orientation and tilt angle of a solar array to achieve optimal performance.

In Chapter 3, we'll explore solar electricity—the types of solar cells on the market today, their efficiencies, and how they generate electricity. I will also introduce you to some new solar electric technologies that are currently being developed.

In Chapter 4, we will explore the feasibility of tapping into solar energy to produce electricity at your site. I'll show you how to assess your electrical energy needs and determine the size of the solar system you'll need to meet them. You'll also learn why it is so important to make your home as energy efficient as possible before you install a solar electric system.

In Chapter 5, we'll examine the types of residential solar electric systems and hybrid systems—for example, wind and PV systems. You will also learn about connecting a PV system to the electric grid and how utilities handle surpluses you may produce, a process called *net metering*. We'll explore ways to expand a solar electric system and ways to add batteries to a grid-tied system.

Chapter 6 introduces inverters, string inverters, microinverters, and optimizers. You'll learn about your options.

In Chapter 7, I'll tackle batteries and the associated equipment required in battery-based solar systems, including charge controllers and backup generators. You will learn about the types of batteries you can install, how to install them correctly, battery maintenance, ways to reduce maintenance, and safety. You will also learn a little bit about sizing a battery bank for a PV system.

In Chapter 8, I'll describe mounting options so you can make the best decisions for your site.

In Chapter 9, we'll explore a range of issues such as permits, covenants, and utility interconnection. I'll discuss whether you should install a system yourself or hire a professional and, if you choose the latter, how to locate a competent installer.

CHAPTER 2

Understanding the Sun and Solar Energy

Things You Need to Know to Size and Mount a Solar Array

The Sun is a massive nuclear reactor camped in the center of our solar system, approximately 93 million miles from Earth. It produces mind-boggling amounts of energy that radiate into space, primarily as light and heat.

As you may recall from the last chapter, only a tiny portion of the Sun's daily output strikes the Earth, warming the planet and powering virtually all life. Every ounce of energy your body needs comes from solar energy. As you may also recall from the last chapter, only a tiny portion of the incoming solar radiation would provide more than enough energy to power our world.

According to the US Department of Energy's National Renewable Energy Laboratory (NREL), to generate the electricity the United States currently consumes, we'd only need to install PVs on 7% of the land surface area now occupied by cities and homes. We could achieve this goal by installing PVs on rooftops of homes, factories, and office buildings and as sunshades over parking lots (Figure 2.1 a–c). We wouldn't have

2.1a

Figure 2.1. *Solar electric systems can be mounted in ways that don't require big structural changes. (a) Quality Copy Services in Union, Missouri, installed modules on a rack south of the building, providing shade in the summer and a great place to park one's car, albeit very carefully, on a hot summer afternoon. (b) The roofs of commercial buildings, like this Pro-lube Express in St. Joseph, Missouri, provide plenty of square footage that can be used to mount large solar arrays. Imagine the size of the solar arrays that could be mounted on large factories and box stores. (c) This array in St. Louis was made to perform double-duty, serving as both a source of electricity and as an awning to provide shade in the summer.*
Credit: Dan Chiras.

to appropriate a single acre of land to make PV our primary source of electricity!

In this chapter, we'll explore the Sun itself—its path through the sky over the course of a year, and the energy it produces. I'll also cover topics that will help you understand ways to get the most out of a solar electric system.

Debunking a Myth

The Sun's intensity varies by region. The Desert Southwest, for instance, is blessed with sunlight. The Midwest receives less but still plenty of sunlight to make a solar electric system worthwhile. On an annual basis, for example, Kansas City receives about 25% less sunlight than Phoenix. Buffalo, New York, one of the cloudiest regions in the United States, receives about 50% less sunlight than Kansas City. But don't be fooled. Solar electric systems work well in *all* regions, although the size of systems must be increased in cloudier regions to make up for reduced sunshine. The common complaint that "there's not enough sunlight where I live," is dead wrong. Don't fall into that trap. Remember: solar systems still produce electricity on overcast days.

Understanding Solar Radiation

The Sun's energy, known as *solar radiation,* can also be referred to as *electromagnetic radiation.* It includes X-rays, ultraviolet radiation, light, and heat. Several other forms of electromagnetic radiation are also released by the Sun, as shown in Figure 2.2.

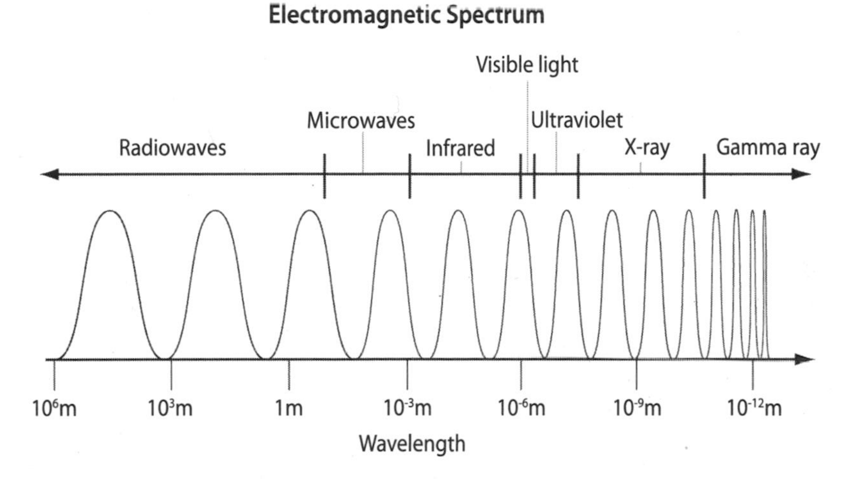

Figure 2.2. *Electromagnetic Spectrum. The Sun produces a wide range of electromagnetic radiation, from gamma rays to radio waves, as shown here. PV cells convert visible light into electric energy, specifically direct current electricity.* Credit: Anil Rao.

Although the Sun releases numerous forms of electromagnetic radiation, most of it (about 40%) consists of infrared radiation (or heat) and visible light (about 55%). Solar electric (PV) modules capture only a portion of this energy—notably, the energy contained in the visible and lower end of the infrared portions of the spectrum. We call the latter *near infrared radiation.*

Electromagnetic radiation from the Sun travels virtually unimpeded through space until it encounters the Earth's atmosphere. Once it enters the atmosphere, the ozone layer, clouds, water vapor, and air pollutants (including dust) either absorb or reflect a large portion of the Sun's energy or radiate it back into space.

Irradiance

The amount of solar energy striking a surface is known as *irradiance.* Irradiance is measured in watts per square meter (W/m^2). Remember, watts is a measure of power production or consumption.

Solar irradiance measured just before the Sun's radiation enters the Earth's atmosphere is about 1,366 W/m^2. On a clear day, nearly 30% of the Sun's radiant energy is absorbed or reflected by dust and water vapor in the Earth's atmosphere. By the time the incoming solar radiation reaches a solar array, the 1,366 W/m^2 measured in outer space has been winnowed down to 1,000 W/m^2. (As you shall soon see, this is the highest irradiation encountered at most locations throughout the world and is called *peak sun.* It's also the intensity of light used to determine the rated power of a solar module.)

Even though the Sun experiences long-term (11-year) cycles, during which its output varies, over the short term, irradiance remains fairly constant. On Earth, however, solar irradiance (the intensity of sunlight) varies during daylight hours at any given site. At night, as you'd expect, solar irradiance is zero. As the Sun rises, irradiance increases slowly but surely, plateauing between 10 a.m. and 2 p.m. Afterward, irradiance slowly decreases, falling once again to zero at night.

Changes in irradiance are determined by the angle at which the Sun's rays strike the surface of the Earth. Because the Sun's position in the sky changes during the day (thanks to the rotation of the Earth on its axis), the angle of incoming solar radiation changes throughout the day.

The angle at which the Sun's rays strike the Earth, in turn, affects two other factors: energy density (Figure 2.3) and the amount of atmosphere through which sunlight must travel to reach the Earth's surface (Figure 2.4).

Let's start with energy density. As shown in Figure 2.3, early morning and late-day sunlight is low-angled sunlight that delivers much less energy per square meter than high-angled sunlight. This results in decreased irradiance. As the Sun makes its way across the sky during the morning, however, the sunlight begins to stream in from directly above, increasing irradiance.

Figure 2.3. *Energy Density varies during the day and by season. Low-angled Sun results in lower energy density than high-angle sunlight.* Credit: Anil Rao.

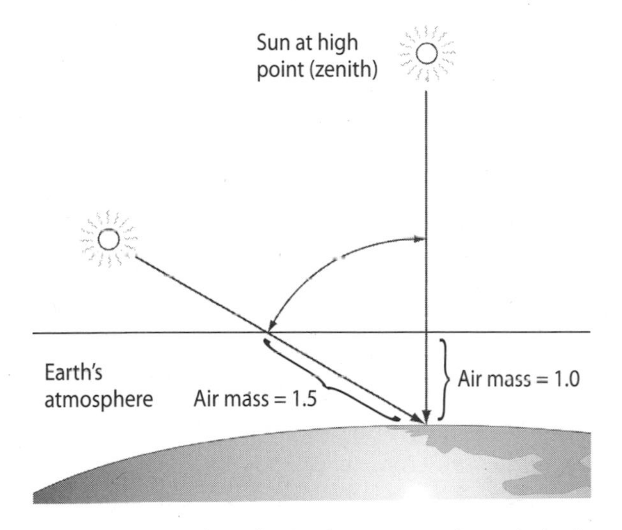

Figure 2.4. *Atmospheric Air Mass and Irradiance Reaching the Earth's Surface. Early and late in the day, sunlight travels through more air in the Earth's atmosphere, which decreases the amount of energy reaching a solar electric array, decreasing its output. Maximum output occurs when the Sun's rays pass through the least amount of atmosphere, at solar noon. Three-quarters of the daily output from a solar array occurs between 9 a.m. to 3 p.m.* Credit: Anil Rao.

The angle at which incoming solar radiation strikes the Earth also affects the amount of atmosphere through which sunlight must travel, as shown in Figure 2.4. The more atmosphere, the more filtering of incoming solar radiation, the less sunlight makes it to Earth, and the lower the irradiance. I'll elaborate on this concept shortly.

Irradiation

Irradiance, as just noted, is a measure of instantaneous power—how many watts are striking the Earth per square meter at any one moment. Although irradiance is an important measurement, what most solar installers need to know is irradiance over time—the amount of energy they can expect to capture each day. Irradiance over a period of time is referred to as *solar irradiation* or simply *irradiation.*

Irradiation is expressed as watts per square meter striking the Earth's surface (or a PV module) for some specified period—for example, per hour. One hundred watts of sunlight striking a square meter of surface for one hour is 100 watt-hours per square meter. Solar radiation of 500 watts of solar energy striking a square meter for an hour is 500 watt-hours per square meter of irradiation. As you shall soon see, solar installers use this number to determine the output of a solar electric system.

To help keep these terms straight, remember that irradiance is an *instantaneous measure of power.* Irradiation, on the other hand, is a *measure of power delivered over a certain amount of time.* It's a quantity that physicists refer as *energy.* To keep the terms—power and energy—separate, I liken irradiance to the speed of a car. Like irradiance, speed is an instantaneous measurement. It tells us how fast a car is moving at any moment. It is a rate. Irradiation is akin to the distance a vehicle travels during a certain amount of time, for instance, an hour. Distance is determined by multiplying the speed of a vehicle by the time it travels at that speed. It is a quantity, not a rate. Irradiation is a quantity as well.

Figure 2.5 illustrates the concepts graphically. As shown, irradiance is the single black line in the graph—the number of watts per square meter at any given time. The area under the curve is the total solar irradiance during a given period, in this case, a single day. The total irradiance in a day is irradiation. Solar irradiation at any prospective solar site is needed to determine the size of the system required to meet the electrical demands of customers.

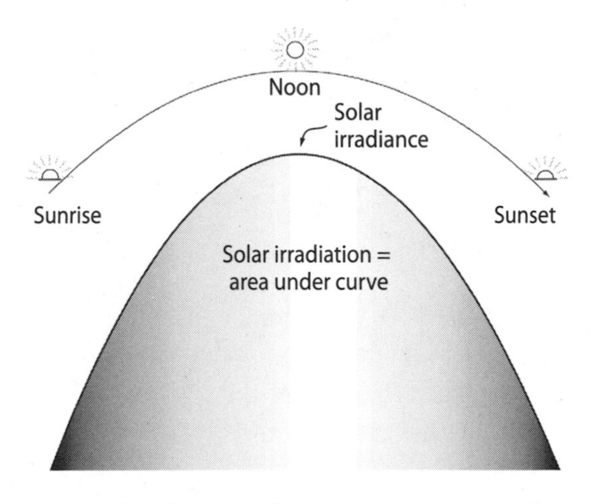

Figure 2.5. *Graph of Solar Irradiance and Solar Radiation. This graph shows both solar irradiance (watts per square meter striking the Earth's surface) and solar irradiation (watts per square meter per day). Note that irradiation is total solar irradiance over some period of time, usually a day. Solar irradiation is the area under the curve.* Credit: Anil Rao.

Peak Sun and Peak Sun Hours

Another measurement installers use when designing PV systems is *peak sun.* As briefly noted, peak sun is the maximum solar irradiance available at most locations on the Earth's surface on a clear day—1,000 W/m^2 (or 1 kWh/m^2). One hour of peak sun is known as a *peak sun hour.* Four hours of peak sun are four peak sun hours. The average number of peak sun hours occurring in various locations during a year varies, being highest in the southwestern US and lowest in the northeast and northwest and Canada.

Average peak sun hours per day are used to compare solar resources of different regions. The annual average in southern New Mexico and Arizona, for instance, is 6.5 to 7 peak sun hours per day. In Missouri, it's 4.7 to 4.9 peak sun hours per day. In cloudy British Columbia, it's about 4. Remember, though, these are annual averages. In the summer in virtually all locations, peak sun hours are considerably greater than in the winter.

Installers determine peak sun hours by consulting detailed state maps, tables, or various web sites. They use average peak sun hours to size solar arrays—that is, to determine how large of an array is required to meet a customer's needs. They also use it to estimate the annual output of existing arrays to troubleshoot—for example, if a solar array appears not to be performing as well as expected.

It's important to note that the average peak sun hours per day for a given location doesn't mean that the Sun shines at peak intensity during that entire period. In fact, peak sun conditions—solar irradiance equal to 1,000 watts/m^2—will very likely only occur two to four hours a day. To learn how peak sun hours is calculated, check out the sidebar on this page.

Calculating Average Daily Peak Sun Hours

Peak sun hours are calculated at a given location by determining the total irradiation (watt-hours per square meter) during daylight hours and dividing that number by 1,000 watts per square meter. On a summer day, for example, let's assume that solar irradiance averages 600 watts per square meter for 4 hours, 800 watts per square meter for 4 hours, and 1,000-watts per square meter for 2 hours.

Four hours at 600 W/m^2 is 2,400 watt-hours/m^2. Four hours at 800 W/m^2 is 3,200 watt-hours/m^2. Two hours at 1,000 W/m^2 is 2,000 watt-hour/m^2. Add them up, and you have 7,600 watt-hours/m^2. That's the daily solar irradiation. The number of peak sun hours is 7,600 watt-hours/m^2 divided by 1,000. In this case, we'd have 7.6 peak sun hours. You can see that even though average peak sun hours is 7.6, solar irradiance only reached 1,000 W/m^2 (peak sun) for 2 hours.

Sun and the Earth: Understanding the Relationships

When designing and installing a PV system, it's important to follow certain rules. These rules will ensure maximum electrical production throughout the year. The first rule is that, for maximum solar output, a solar array should be oriented to true south (in the Northern Hemisphere), which is explained shortly. The modules must be angled properly to ensure maximum energy production, too. An understanding of the Earth's relationship to the Sun as it travels through its orbit will help you understand exactly what's required.

Day Length and Altitude Angle: The Earth's Tilt and Orbit Around the Sun

As all readers know, the Earth orbits around the Sun, completing its sojourn every 365 days. As shown in Figure 2.6, the Earth's orbit is elliptical. As a result, the distance from the Earth to the Sun varies throughout the year. Surprisingly, the Earth is closest to the Sun in the Winter, and farthest away in the summer.

As shown in Figure 2.6, the Earth's axis is tilted 23.5°. The Earth maintains this angle throughout the year as it orbits around the Sun. Look carefully at Figure 2.7 to see that the angle remains fixed—almost as if the Earth were attached to a wire anchored to a fixed point in space.

Because the Earth's tilt remains constant, the Northern Hemisphere is tilted away from the Sun during the winter. As a result, the Sun's rays pass through Earth's atmosphere at a very low angle. This reduces energy density and increases the amount of sunlight absorbed by the atmosphere. This is one reason winters are colder, and it is also why solar arrays produce less electricity in the winter.

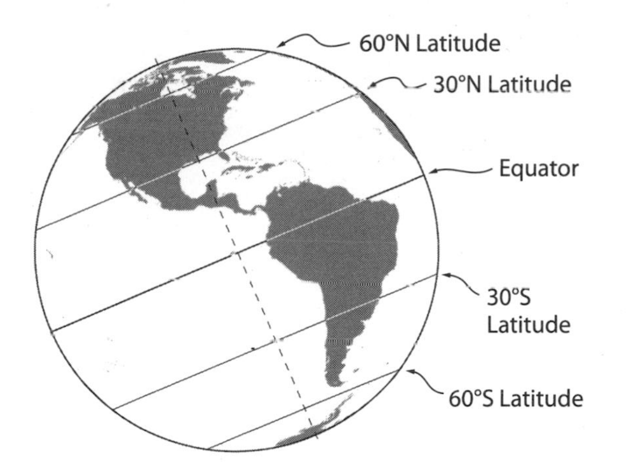

Figure 2.6. *The Earth's Axis. The Earth is tilted on its axis of rotation, a simple fact with profound implications on solar energy use on Earth.* Credit: Anil Rao.

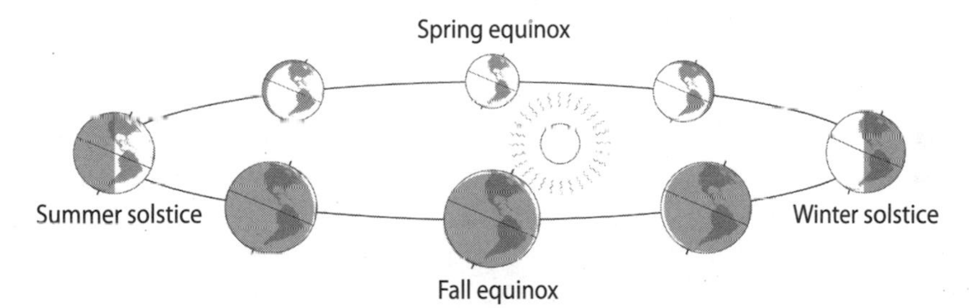

Figure 2.7. *The Earth's Orbit around the Sun. As any school child can tell you, the Earth orbits around the Sun. Its orbit is not circular, but elliptical, as shown here. Notice that the Earth is tilted on its axis and is farthest from the Sun during the summer. Because the Northern Hemisphere is angled toward the Sun, however, our summers are warm.* Credit: Anil Rao.

Solar gain is also reduced in the winter because days are shorter—that is, there are fewer hours of daylight during winter months. All three factors combine to reduce the amount of solar energy available to a PV system in the winter.

In the summer in the Northern Hemisphere, the Earth is tilted *toward* the Sun, as shown in Figures 2.6 and 2.8. This results in several key changes. One of them is that the Sun is positioned higher in the sky. Sunlight shines in at a steeper angle. This, in turn, increases energy density and reduces absorption and scattering of incoming solar radiation. Both factors increase solar irradiance and irradiation. Increased irradiation increases the output of a PV array—the energy production. Because days are longer during the summer, the output of a solar array is also a bit higher during this time of year.

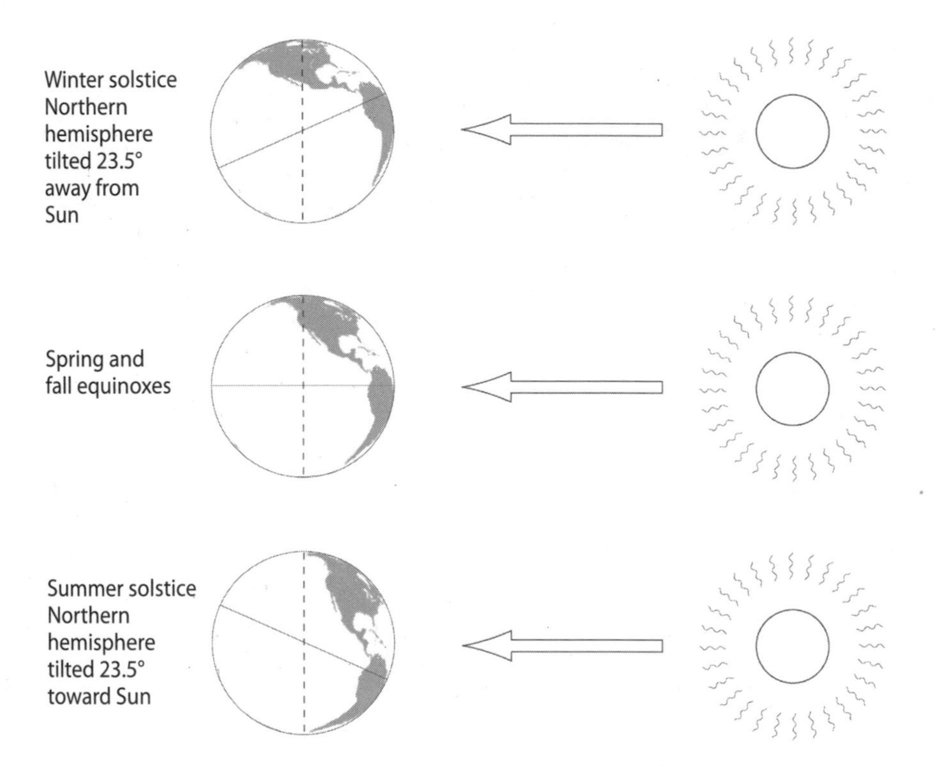

Figure 2.8. *Summer and Winter Solstice. Notice that the Northern Hemisphere is bathed in sunlight during the summer solstice because the Earth is tilted toward the Sun. The Northern Hemisphere is tilted away from the Sun during the winter.* Credit: Anil Rao.

Figure 2.9 shows the position of the Sun as it "moves" across the sky during different times of the year. It's a result of the changing relationship between the Earth and the Sun. As shown, the Sun "carves" a high path across the summer sky. It reaches its highest arc on June 21, the longest day of the year, also known as the *summer solstice.* (*Sol* stands for sun, and *stice* is derived from a Latin word that means to stop.)

Figure 2.9 also shows that the Sun "carves" a low path across the sky in the winter. The lowest arc occurs on December 21, the shortest day of the year. This is the *winter solstice.*

The angle between the Sun and the horizon at any time during the day is referred to as the *altitude angle.* As shown in Figure 2.9, the altitude angle decreases from the summer solstice to the winter solstice. After the winter solstice, however, the altitude angle increases, growing a little each day, until the summer solstice. Day length changes along with altitude angle, decreasing about two minutes a day for six months from the summer solstice to the winter solstice, then increasing two minutes a day until the summer solstice arrives once again.

The midpoints in the six-month cycles between the summer and winter solstices are known as *equinoxes.* (The word *equinox* is derived from the Latin words *aequus* [equal] and *nox* [night].) On the equinoxes, the hours of daylight are nearly equal to the hours of darkness. The spring equinox occurs around March 20 and the fall equinox occurs

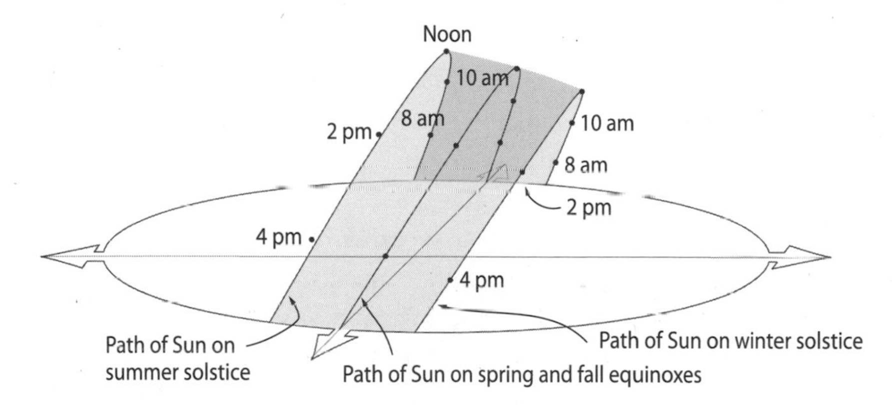

Figure 2.9. *Solar Paths. This drawing shows the position of the Sun in the sky during the day on the summer and winter solstices and the spring and fall equinoxes. This plot shows the solar window—the area you want to keep unshaded so the Sun is available for generating solar electricity.* Credit: Anil Rao.

around September 22. These dates mark the beginning of spring and fall, respectively.

The altitude angle of the Sun changes day by day as it moves to and from the solstices, but it also varies according to the time of day. This change in altitude angle is also determined by the rotation of the Earth on its axis. As seen in Figure 2.9, the altitude angle increases between sunrise and noon, then decreases to zero once again at sunset.

In addition to the change in the altitude angle of the Sun, the Sun's position in the sky relative to the points on the compass changes during the day. (This is referred to as the *azimuth angle.*) You can verify this yourself, by simply going outside on a sunny day with a compass and viewing the sun from the same fixed point every couple of hours. You'll see that the Sun's position changes in relation to your position. If you were dedicated to the project and checked the Sun's position at the solstices and equinoxes, you'd be able to create a sun path diagram like the one in Figure 2.9.

Implications of Sun-Earth Relationships on Solar Installations

Solar arrays are typically installed either on the top of poles secured to the ground or, more commonly, on racks that are mounted on the ground or on roofs or the sides of buildings. Installers are careful to orient the array properly and tilt the rack at the proper angle. Doing so ensures maximum solar gain and electrical production.

For best year-round production, a solar array should be mounted on a rack that can track the Sun across the sky from sunrise to sunset. For slightly better performance, a tracking array should also change its tilt angle (described shortly) so that the array adjusts for changes in the altitude angle and is therefore always oriented so that the Sun's rays strike it at a 90° angle.

Unfortunately, tracking arrays can be expensive and may not be worth the money. Tracking arrays are also not installed on roofs, where most residential systems are placed. In such instances, installers typically install fixed arrays—that is, arrays that remain in a fixed position year round. They are installed at a certain tilt angle and are oriented in the same direction year round. So how should arrays be oriented, and at what angle should the modules be tilted?

For best year-round performance, fixed arrays should be oriented to true south (described shortly). The angle at which they are installed—the tilt angle—should be set so the modules absorb the most sunlight throughout the year.

As shown in Figure 2.10, the tilt angle is the angle between the surface of the array and an imaginary horizontal line extending back from the bottom of the array. The ideal tilt angle that most sources recommend is equal to the latitude of the site. If you live at 45° north latitude, for instance, most installers suggest tilting the array at a 45° angle. In my experience, however, I've found that for best year-round production, the tilt angle of a fixed array should be latitude minus 5°. This recommendation is based on data supplied by NASA's website "Surface Meteorology and Solar Energy" (eosweb.larc.nasa.gov).

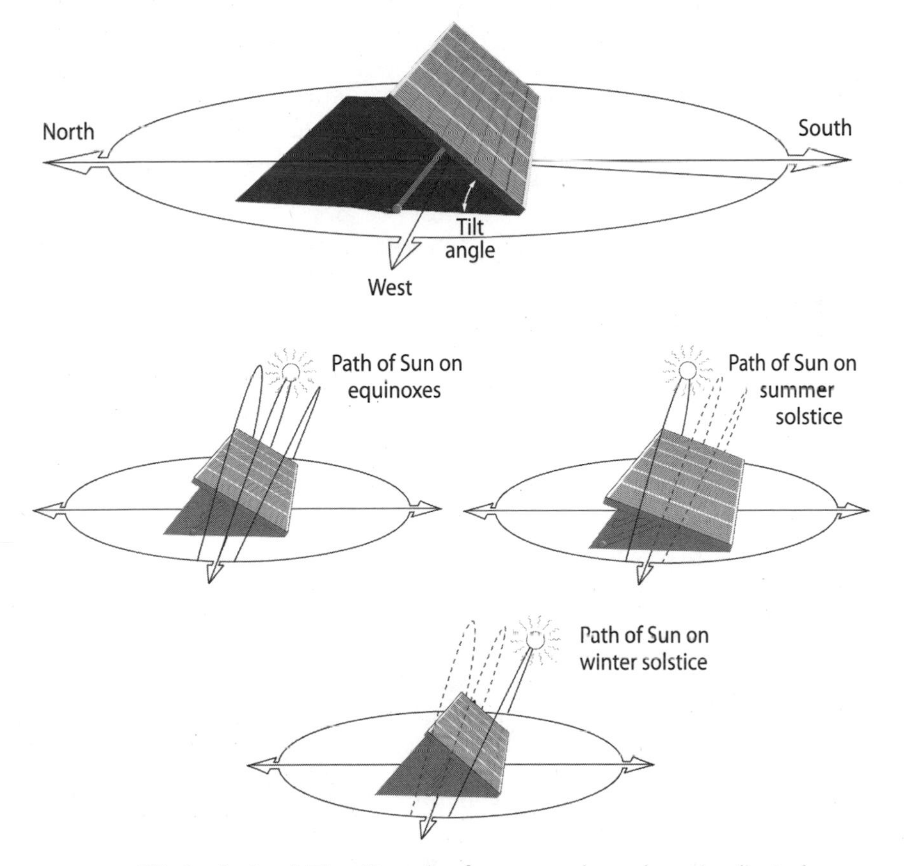

Figure 2.10. *Tilt Angle. (top) The tilt angle of an array, shown here, is adjusted according to the altitude angle of the Sun. The tilt angle can be set at one angle year round, known as the optimum angle, or (bottom) adjusted seasonally or even monthly, although most homeowners find this too much trouble.* Credit: Anil Rao.

When mounting PV arrays, always be sure to orient them to true south if you live in the Northern Hemisphere and true north if you live in the Southern Hemisphere. True north and south are imaginary lines that run from the North Pole to the South Pole, parallel to the lines of longitude (True north and south are also known as true geographic north and south.) They're not to be confused with magnetic north and south.

Magnetic north and south are lines of force created by the Earth's magnetic field. They are measured by compasses. Unfortunately, magnetic north and south rarely line up with the lines of longitude—that is, they rarely run true north and south. In some areas, magnetic north and south can deviate quite significantly from true north and south. How far magnetic north and south deviate from true north and south is known as the *magnetic declination.*

Figure 2.11 shows the deviation of magnetic north and south—the magnetic declination—from true north and south in North America. You may want to take a moment to study the map. Start by locating your state and then reading the value of the closest isobar.

If you live in the eastern United States, along the -10° bar, magnetic south is ten degrees off true south. Notice that the lines of magnetic deviation in the eastern half of the country are labeled with negative signs. This is a shorthand designation that indicates a westerly declination. Translated that means that true south is 10° west of magnetic south. If you are standing at a site located along this line, your compass will point to magnetic south, but true south is 10° west of magnetic south.

If you live in the midwestern and western United States, the lines have no sign. This indicates an easterly declination. That is, true south lies east of magnetic south. When standing anywhere along the 15° line, for instance, true south will be 15° east of magnetic south.

To determine the magnetic declination at your site, it is best to contact a local surveyor or a nearby airport. Because magnetic fields can change over a period of a few years, you need a reliable, up-to-date source. Be sure to ask whether the magnetic declination is eastly or westerly. Also bear in mind that compass readings at a site may deviate because of local iron ore deposits, metal buildings, or even vehicles. So, if you take a magnetic reading too close to a vehicle or a metal barn, the compass may not point precisely to magnetic south. In some cases, it could be way off. Local variations in the magnetic field are called *magnetic deviation.* To avoid magnetic

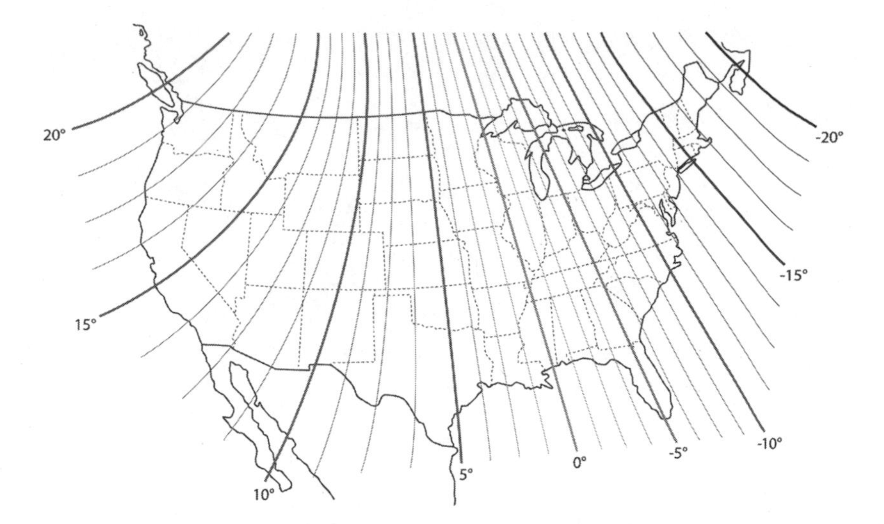

Figure 2.11. *Map of Magnetic Declination. The isobars on this map indicate magnetic declination in the United States. Note that negative numbers indicate a west declination (meaning true south lies west of magnetic south). Positive numbers indicate an east declination (meaning true south is east of magnetic south).* Credit: Anil Rao.

deviation, be sure to move away from metal buildings, metal fences, vehicles, and other large metal objects.

If you think that your site may be compromised by a local deposit of iron ore, a failsafe method of finding true north and south is to head outside on a cloudless night and locate the North Star, Polaris. "The North Star is always either exactly or just a few tenths of a degree from true north," says Grey Chisholm, author of "Finding True South the Easy Way" in *Home Power* (Issue 120). How do you locate it?

The North Star is situated at the end of the handle of the Little Dipper (Figure 2.12). I've found that the Little Dipper can be a bit difficult to discern. To locate the North Star, follow "the two pointer stars on the outside of the Big Dipper's cup," says Chisholm. A line drawn from them will take you directly to the North Star.

To determine true north and south from the North Star, drive a fence post into the ground, then move south a bit so you are in line with the post and the North Star. Drive a second fence post into the ground at this point. You've now created a line that runs true north and south. Use this line to orient your PV array.

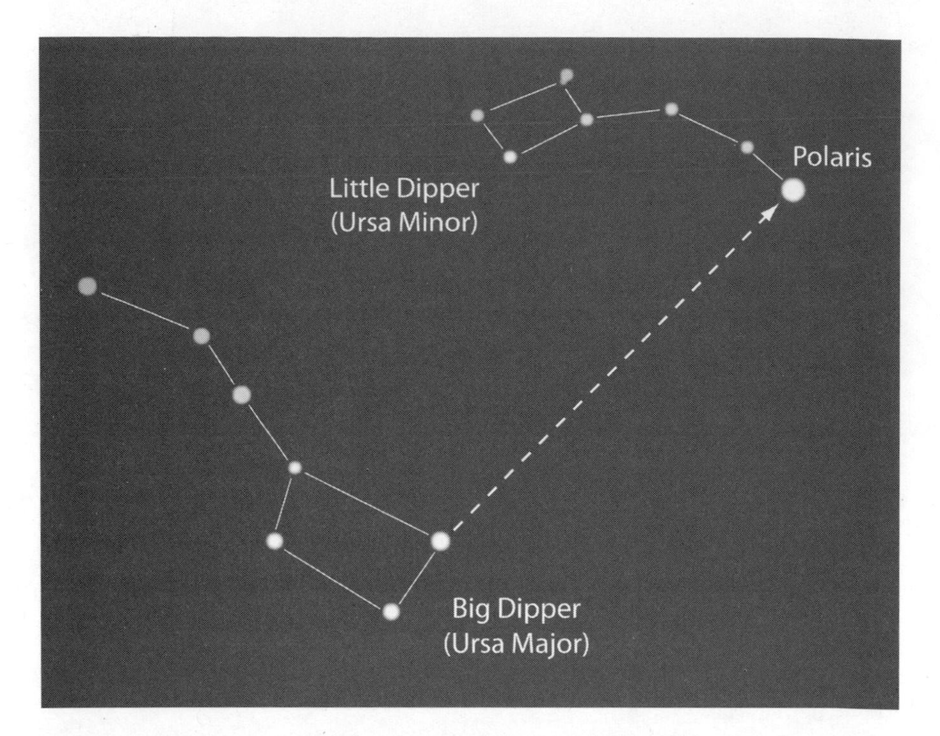

Figure 2.12. *North Star and Little Dipper. Use the North Star (Polaris) to locate true north.* Credit: Anil Rao.

Conclusion

Solar energy is an enormous resource that could contribute mightily in years to come. To make the most of it, though, a solar electric system has to be installed properly—that is, oriented properly to take into account the Sun's daily and seasonally changing position in the sky. In the next chapter, we turn our attention to the technologies that capture solar energy: solar cells and solar modules.

CHAPTER 3

Understanding Solar Electricity

Types of Modules, How They Are Rated, and Advances in Solar Technology

Many people think that solar electricity is a new technology. Or, they may view it as a by-product of the US space program's efforts to power satellites—work that dates back to the early 1950s. Fact is, the development of solar electricity began well over 100 years ago. In recent years, however, solar electricity has evolved from a scientific curiosity to a full-fledged source of electricity throughout the world.

In this chapter, I'll describe features of modern PV modules and describe how they convert sunlight energy into electricity. I'll discuss the types of modules on the market today and some of the newest and most promising PV technologies. I'll also introduce you to some terminology that will be helpful to you if you want to install your own system or hire someone to do the job for you. Remember: knowledge is power. In this case, clean, reliable, and affordable power!

What Is a PV Cell?

Photovoltaic cells are solid-state electronic devices like the computer chips in our laptops, tablets, cell phones, and smart TVs. PV cells and computer chips are referred to as *solid-state devices* because they are made of a solid material: silicon. Although researchers are exploring the use of other materials to make solar cells, most cells in use today are made from silicon. Silicon is extracted from silicon dioxide, the most abundant mineral in the Earth's crust. Most silicon is extracted from silica sand—the white-as-sugar sand you find on some beaches.

Silicon atoms contain electrons that orbit around a central nucleus that contains protons and neutrons. In silicon, however, some of the electrons can be knocked out of their orbit when struck by sunlight. These loose electrons can be made to flow together, creating an electrical current.

Most solar cells in use today are thin wafers of silicon about 1/100th of an inch thick. As shown in Figure 3.1, most solar cells consist of two layers—a very thin upper layer and a much thicker lower layer. The upper layer is made of silicon and phosphorus atoms; the bottom layer consists of silicon and boron atoms.

When sunlight strikes the silicon atoms in solar cells, it boots electrons out of orbit. These electrons flow toward the surface where they are picked up by thin metal contacts on the surface—or just below the surface—of the solar cells. Because the solar cells are wired in series in a PV module, they form a continuous circuit. Consequently, electrons released from one cell flow to the next cell, and then to the next one, etc., until they reach the negative terminal of the PV module. (That's where a wire conducts electricity out of a module.)

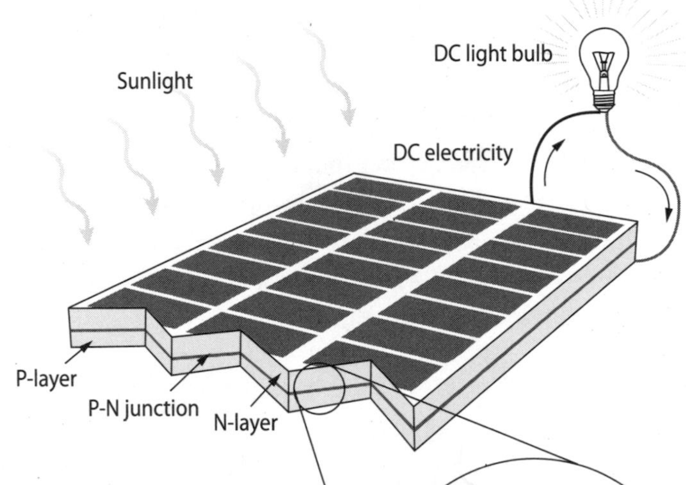

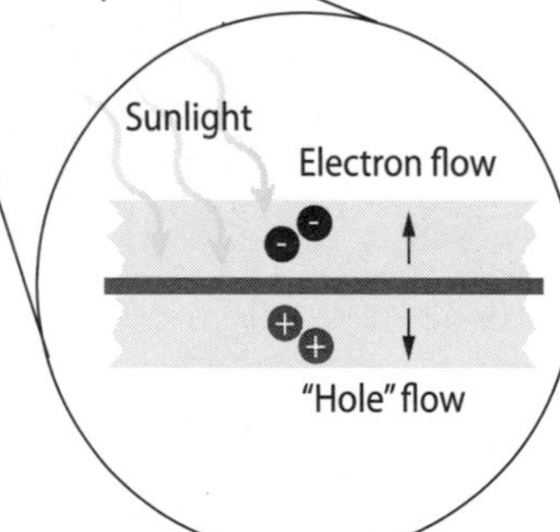

Figure 3.1. *Solar Cell Powering DC Light. Electrons are ejected from atoms in the solar cell. These electrons are energized by photons of light striking the solar cell. The energy they gain in this process is used to power various loads, like light bulbs. The n- and p-layers are described in the text.* Credit: Forrest Chiras.

Solar-energized electrons then flow from the array to an inverter, which converts the DC electricity to AC. After delivering the energy they gained from sunlight to the inverter, the de-energized electrons return through another wire to the solar array. The electrons then flow back into the solar cells, filling the empty spots ("holes") left in the atoms created by their ejection. This permits the flow of electrons to continue *ad infinitum.* For a more detailed description, you may want to check out my book, *Power from the Sun.*

Types of PV

Solar cells can be made from a variety of materials. By far the most common is silicon. Even though there are other materials that are more efficient at converting sunlight energy to electrical energy, silicon semiconductors dominate the market because they currently produce the most electricity at the lowest cost. As you shall soon see, three forms of silicon are found in solar modules: monocrystalline, polycrystalline, and amorphous.

Monocrystalline PVs

Monocrystalline cells—aka *single-crystal cells*—were the very first commercially manufactured solar cells. They've been around since the early 1950s. Monocrystalline cells are made from ultrathin wafers sliced from a single crystal of pure silicon known as a *silicon ingot* (Figure 3.2).

Figure 3.2. *Silicon Ingot. This ingot is a huge crystal of silicon that is sliced to make monocrystalline PV cells.*
Credit: SolarWorld.

The long, cylindrical single-crystal ingots are made by melting highly purified chunks of silicon and a small amount of boron. Once the raw material is melted, a seed crystal is dipped in the molten material, known as the *melt.* This crystal is rotated and slowly withdrawn from the melt. As it's withdrawn, silicon and boron atoms attach to the seed, duplicating its crystal structure. As the crystal is withdrawn, it grows larger and larger. Eventually, the ingot may grow to a length of 40 inches (100 cm) and a diameter of 6 to 8 inches (15 to 20 cm).

Once fully extracted, the ingot is cooled. Manufacturers then trim the round ingot to produce a nearly square ingot—actually a square ingot with rounded corners. It is then sliced with a diamond wire saw, producing ultrathin wafers (Figure 3.3). Waste from this process is remelted and reused—often to make the next type of solar cell, polycrystalline.

Trimming an ingot prior to slicing allows manufactures to place more solar cells in a module. This, in turn, increases the output per square centimeter. (You can fit more nearly square solar cells in a module than round cells.) Because monocrystalline cells are "squared off," modules made from them can be easily identified, even from a distance, by the white space at the juncture of four cells (Figures 3.4 and 3.5).

Monocrystalline cells boast the highest efficiency of all conventional PV cells, with commercially available cells attaining efficiencies in the

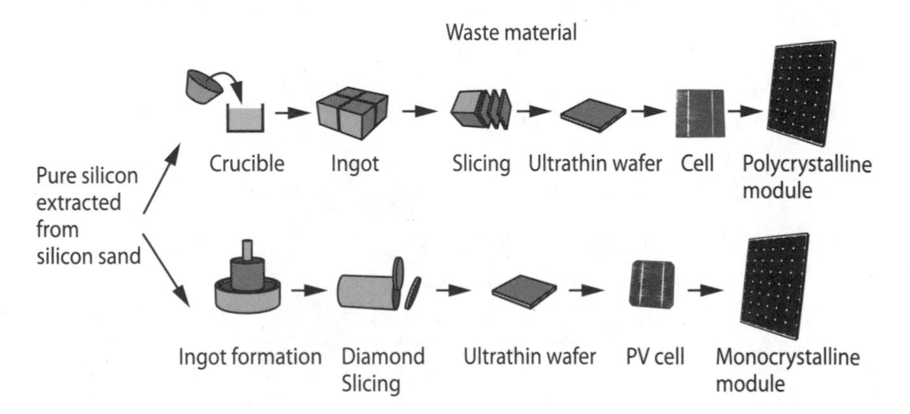

Figure 3.3. *Steps in Wafer Production. As shown here, silica is used to produce high-purity silicon, which is then used to make monocrystalline and polycrystalline solar electric cells. Note that waste from the production of monocrystalline solar cells is also used in the production of polycrystalline solar cells.*
Credit: Forrest Chiras.

range of 15% to 20%. Efficiency is a measure of energy input vs. energy output. If a module is 16% efficient, it will produce 16 units of electricity for every 100 units of sunlight energy. Although you may occasionally read about even-higher efficiencies, these reports are typically about solar cells still in development and much too costly for commercial production.

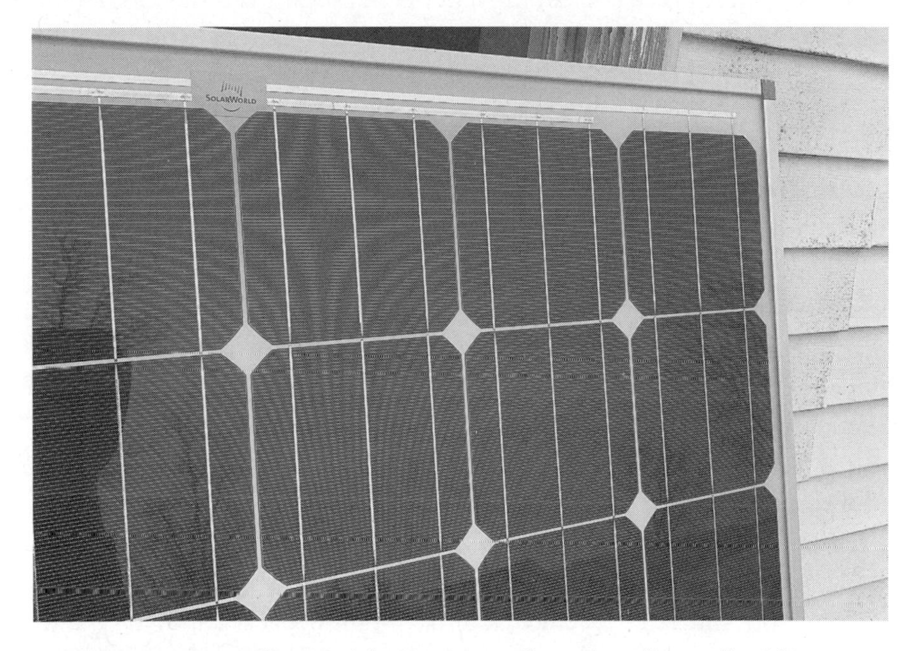

Figure 3.4. *Monocrystalline Module. Monocrystalline solar cells are sliced from a long cylindrical ingot of silicon doped with boron. You can identify the type of solar cell by its distinct shape, as described in the text.* Credit: Dan Chiras.

Figure 3.5. *Monocrystalline Module. Even from a distance you can tell that these solar modules are made from monocrystalline solar cells.* Credit: Dan Chiras.

Polycrystalline PV Cells

Polycrystalline solar cells emerged on the scene in 1981. Just like monos, they are made from pure silicon with a trace of boron. To make a polycrystalline cell, however, the molten material is poured into a square mold. It is then allowed to cool very slowly. As the ingot cools, many smaller crystals form internally. Once cooled, the ingot is removed from the mold, then sliced using a diamond wire saw, creating wafers used to fabricate solar cells, as shown in Figure 3.3.

Polycrystalline solar cells in previous years (in the 1990s, for example) contained many larger crystals, like the ones shown in Figure 3.6a. Today, however, silicon crystals in polycrystalline cells are often so small that they are undetectable to the naked eye (Figure 3.6b).

Modules with polycrystalline solar cells are slightly less efficient than monos—on average, they are about 15% to 17% efficient. The reason for this is that the boundaries between these crystals in a solar cell reduce the flow of electrons through the cell, reducing their output. Even though polycrystalline solar cells are slightly less efficient, they require much less energy to manufacture. As a result, they are cheaper to make. Manufacturers, in turn, pass the savings on to customers. In my experience, polys are typically about 15% cheaper than monos.

Modules made from polycrystalline cells have a distinctive appearance. The cells are square or rectangular and are tightly packed into a module. (You won't see any open space at the junction of solar cells.)

Although polycrystalline modules are slightly less efficient, it is important to note that tight packing makes up for their lower efficiency. As a result of the higher packing density and slightly larger module size, a 285-watt module made from polycrystalline PV cells will produce the same amount of electricity as a more costly 285-watt module made from slightly more efficient monocrystalline PV cells.

Another factor to consider when shopping for modules, according to some sources, is that polycrystalline modules perform a little more efficiently than monocrystalline modules when hot—as they are when basking in bright sunlight. Because of these factors, I advise prospective solar clients to avoid the must-have-the-most efficient—that is, monocrystalline—module mentality. Check out polycrystalline modules—they could save you a substantial amount of money. Don't assume that you are better off buying monocrystalline modules just because they have a slightly higher efficiency.

Figure 3.6. (a) *Close-up of Older Polycrystalline Cell Showing Crystals.* (b) *Newer Polycrystalline Cell. Older polycrystalline solar cells contained very large crystals. Newer polycrystalline solar cells contain very small crystals which are extremely difficult to see.* Credit: Dan Chiras.

Adding the N-Layer and Building the PV Cell

Monocrystalline and polycrystalline wafers sliced from ingots contain silicon and a tiny amount of boron. Newly produced wafers constitute the p-layer, the thicker, bottom layer of PV cells. How is the much thinner, phosphorus-containing n-layer formed?

After the p-layer has been created, the wafer is dipped in a solution of sodium hydroxide; this removes impurities deposited on the wafers during manufacturing. It also etches (roughens) the surfaces of the wafer. Etching also reduces reflection, improving the efficiency of solar cells. (Reduced reflection means more sunlight is absorbed by the cell.)

Once etched, the wafers are placed in a diffusion furnace and heated. Phosphorous gas is introduced into the furnace. In this high-temperature environment, phosphorus atoms penetrate the exposed top surface. This results in the formation of a very thin superficial n-type layer (Figure 3.7). The junction between these two layers is known, quite appropriately, as the p-n junction.

After a wafer is removed from the diffusion furnace, metal contacts are screen-printed onto the face (the n-layer side) of it using a silver paste. The silver paste is then baked on. Contacts are applied in a grid pattern consisting of two or more wide main strips and numerous ultrafine (hair-thin) strips that attach perpendicularly to them (Figure 3.8). This grid collects the electrons that move to the surface of the solar cell when silicon atoms are struck by sunlight.

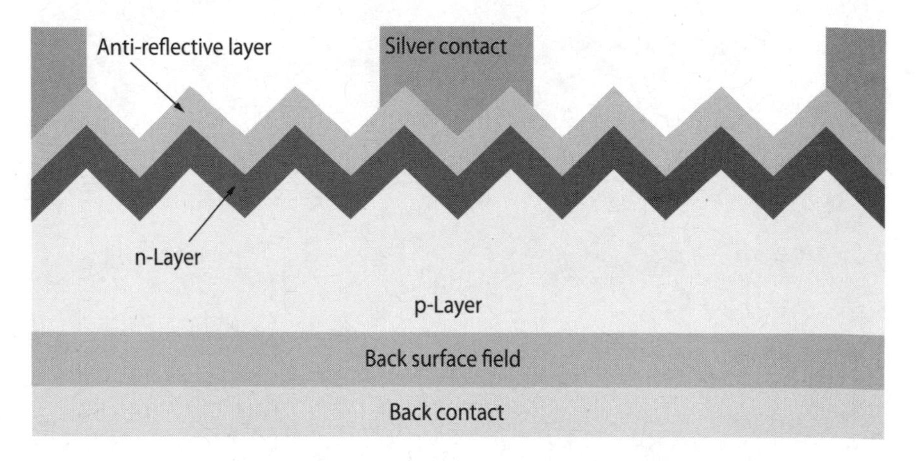

Figure 3.7. *Cross Section through PV. This drawing shows details of the many layers of a traditional mono and poly silicon PV cells.* Credit: Forrest Chiras.

Because silicon is highly reflective, an anti-reflective coating (ARC) is applied to the surface of each solar cell. This black or blue-black coating reduces the reflection of light by 35%, resulting in much greater absorption of the Sun's energy and greater output.

As shown in Figure 3.7, another layer is added to the cell. It is called the *back surface field.* The back surface field improves the efficiency of solar cells. Next comes a thin metal back sheet known as the *back contact.*

Once completed, the cells are individually tested for voltage and amperage (current output) under controlled conditions using artificial light. Because there's a slight variation in voltage among cells, the cells are then sorted according to their voltage. Each solar module is assembled using cells with similar voltages. (That's why a manufacturer will produce

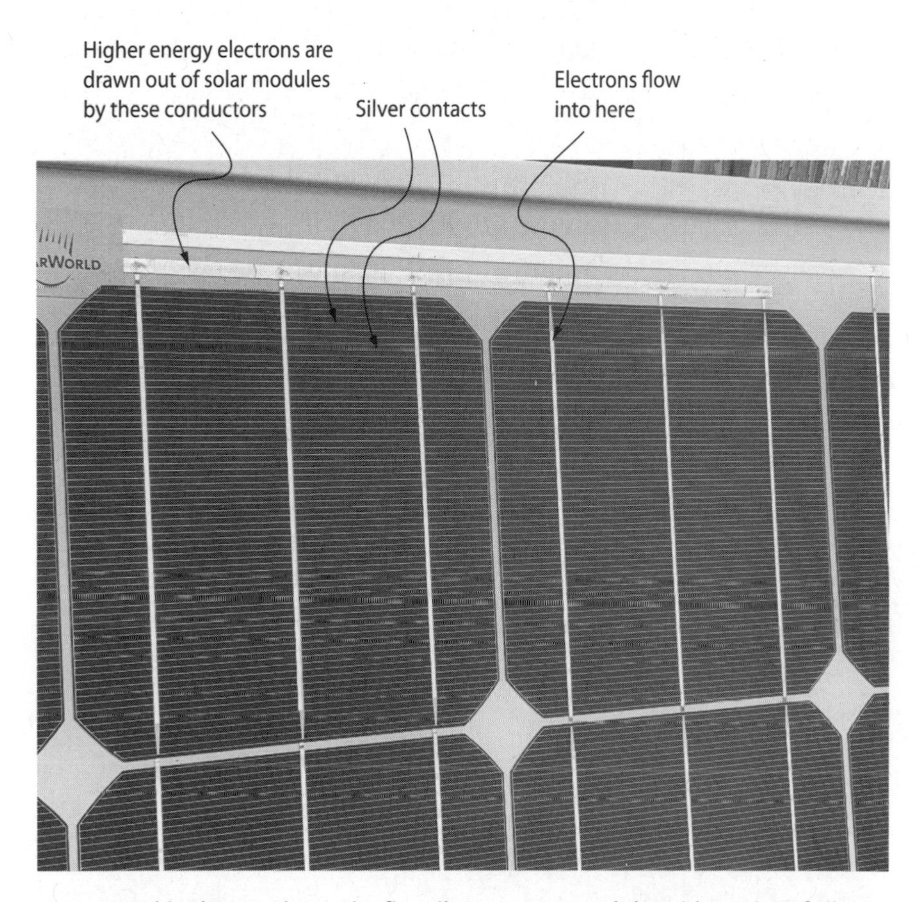

Figure 3.8. *This closeup shows the fine silver contacts and the wider strips of silver to which they connect.* Credit: Dan Chiras.

275-watt modules, 280-watt modules, and perhaps even 285-watt modules. They're from matching cells.)

After being sorted, solar cells are assembled in a module and robotically soldered in series (positive to negative) creating series strings. (This boosts the module voltage.) The surface of the cells is then coated by a thin sheet of clear plastic, known as EVA (ethylene vinyl acetate). Glass is applied to the front of the module and another thin layer of plastic (usually white) is attached to the back, creating a watertight seal. In some modules, glass is substituted for the back plastic. This creates a more translucent module that allows some light to pass through. These modules are ideal for the roofs of carports, patios, bus stops, and other outdoor structures where partial shading is desired.

Virtually all solar modules on the market today come with aluminum frames; they provide rigidity and are used to attach modules to racks. These frames are anodized to prevent oxidation.

Energy Payback for PVs

You may have heard people say that it takes more energy to make a PV system than you get out of it over its lifetime. Fortunately, that's not even close to being accurate.

While it takes energy to make solar cells, modules, and the remaining components of a PV system, the energy payback is amazingly short—only one to two years, according to a study released in 2006 by CrystalClear, a research and development project on advanced industrial crystalline silicon PV technology. The study was funded by a consortium of European PV manufacturers. As Justine Sanchez notes in her 2008 article in *Home Power,* "PV Energy Payback" (Issue 127): "Given that a PV system will continue to produce electricity for 30 years or more, a PV system's lifetime production will far exceed the energy it took to produce it."

CrystalClear's research showed that it takes two years for a PV system with monocrystalline solar cells to make as much energy as was required to manufacture the *entire* PV system. The researchers also calculated that it took 1.7 years for a polycrystalline PV system to produce as much energy as was required to make them. According to the National Renewable Energy

Laboratory (NREL), most thin-film modules require even less energy to produce, so they achieve energy payback in 1 year. Some have even shorter paybacks. Thin-film modules made from cadmium telluride, for instance, have an energy payback of 8 months in ideal conditions.

These energy payback studies were performed in sunlight conditions similar to those found in southern Europe, which has an average insolation of 4.7 peak sun hours. That's the same as in eastern Missouri and western Illinois. For those who live in sunnier climates, the energy payback would be even quicker. For those who live in less-sunny regions, the energy payback would be slower.

Most of the energy required to make a PV system goes into producing the modules—about 93% of the entire energy budget.

Thin-Film Technology: Second-Generation Solar Electricity

In an effort to produce less expensive solar modules, researchers developed a technology, known as *thin-film*. They represent the second generation of solar electricity.

Unlike manufacturers using earlier technologies, thin-film producers manufacture entire modules rather than individual cells that are later assembled into modules. Skipping the step of making and assembling cells into modules saves time, energy, and money.

Most thin-film PVs are made from silicon. In silicon-based thin-film solar designs, known as *amorphous silicon*, silicon is deposited directly onto a metal (aluminum), glass, or plastic backing by a technique known as *chemical vapor deposition*. This creates a thin film of noncrystalline photoreactive material, referred to as an *amorphous* layer. Once it has solidified, a laser is used to delineate cells and create connections between the newly formed cells. Three other semiconductor materials are also used to make thin-film solar modules: cadmium telluride (CdTe); copper indium gallium diselenide (CIGS); and gallium arsenide (GaAs).

Manufacturers have found that they can improve the efficiency of thin-film modules by applying two or more layers of *different* semiconductor materials, each with a unique photo absorption characteristic. This technique results in the absorption of more energy from the incoming solar radiation, which increases output (efficiency).

Multiple-layer cells are called *multi-junction cells.* They are the most efficient cells produced in the world. Two- and three-junction cells boast efficiencies of slightly over 33% and slightly more than 40%, respectively. Bear in mind, however, that these are the products of research laboratories. They are too expensive to manufacture for most applications. You will find high-efficiency multi-junction modules used in outer space to power satellites, an application where increased efficiency is worth the higher price tag. Multi-junction cells are also being used to manufacture concentrated PVs. These are commercial PV arrays with lenses that concentrate light on the PV material, greatly boosting their efficiency.

Thin-film solar offers many advantages. One of them is it can be manufactured in long, continuous rolls. These can be applied to standing seam metal roofs. Short pieces of laminated PV materials can be sewn onto backpacks and gym bags and used to charge cell phones and other portable electronic devices. Thin-film is also used by the military to power portable communication devices. Some thin-film products can even be applied to plate glass windows and skylights to generate electricity. The vast majority of thin-film solar, however, is sandwiched between two layers of glass to create modules—albeit rather heavy ones. (They are about twice as heavy as standard polys and monos.)

Another advantage of thin-film is that it requires considerably less energy to produce. Less energy is required because their manufacture eliminates the costly and energy-intensive ingot production and wafer slicing required to produce mono- and polycrystalline PV cells. Another advantage of thin-film solar is that it uses considerably less semiconductor material.

Yet another advantage of amorphous thin-film PV is that it is less sensitive to high temperatures. At 100°F, a crystalline module will experience a 6% loss in production, while a thin-film amorphous silicon module will experience only a 2% loss. This makes thin-film a good choice for hot, sunny climates like Florida and the tropics.

One key advantage of amorphous silicon thin-film is that it is more shade tolerant than crystalline PV. As Erika Weliczko points out in an article in *Home Power* magazine, "many thin-film cells are as long as the module itself, so shading an entire cell is more difficult than traditional 5- or 6-inch square or round crystalline cells." Shading reduces the amount of sunlight falling on a module, and it also increases electrical resistance inside PV cells. Resistance, in turn, reduces the flow of electricity through a module, decreasing its output.

For years, the chief disadvantage of thin-film PVs has been their low efficiency—in the 6% to 9% range. An array of thin-film modules, therefore, required about twice as much roof space as an array of crystalline silicon modules. Today, the efficiency of thin-film modules for residential installations has increased to be in the range of 11% to 14%, so this is less of a concern.

In today's market, thin-film modules represent a tiny portion of total PV production; most manufacturers produce higher-efficiency monocrystalline and polycrystalline modes. Therefore, thin-film probably won't be an option for most readers.

Rating PV Modules and Sizing PV Systems

PV manufacturers list various parameters of their products on promotional material and spec sheets to assist buyers. These include rated power, power tolerance, efficiency, and maximum system voltage. You need to know what these terms mean, so let's take a look at each one.

Rated power is the wattage a module produces under standard test conditions (STC). It is used to compare one module to another, for example, based on price. It is also useful when sizing a PV array. Installers, for instance, first determine electrical demand of a home, then they determine the size of the PV system needed to meet a customer's needs. Residents of a home, for example, might consume 12,000 kilowatt-hours a year. In eastern Missouri, that would require a 9,000-watt or 9-kW array. If you were installing 300-watt modules, you'd need 30 of them.

The next measurement, *power tolerance,* is the range within which a module either overperforms or underperforms its rated power at STC. It is typically expressed as a percentage. Rated power tolerance ranges from -5% to +5%, depending on the manufacturer, although most modules today boast rated tolerances of -0/+5%. A 300-watt module with a power tolerance of -0/+5% is guaranteed by the manufacturer to produce 300 to 315 watts under STC.

Module efficiency is also a handy number. As you have learned, efficiency is the ratio of output power (watts/m^2) from a module to input power (watts/m^2) from the Sun. An 18% module efficiency means that 18% of the incident (incoming) solar radiation is converted into electricity.

PV Degradation

Thin-film and crystalline PV modules degrade over time, approximately 0.5% to 1.0% per year. According to some sources, degradation occurs because of chemical reactions between boron and oxygen when exposed to sunlight. Other sources pin the decrease in output on degradation of the thin layer of clear plastic coating on the surface of the solar cells, just beneath the glass layer. SunPower, a major US manufacturer of solar modules, claims that degradation occurs primarily as a result of cracking of silicon due to expansion and contraction caused by exposure to hot and cold temperatures. Corrosion of the silver metal contacts may also result in deterioration over time.

Whatever the cause or causes, manufacturers provide a production warranty with their products (in addition to a materials warranty). Production warranties are fairly standard in the industry. Every manufacturer I know of (except for SunPower) guarantees that their modules will still be producing 90% of their minimum peak power in 10 years and 80% in 25 years. SunPower installs a fine copper mesh in their solar cells to reduce cracking. As a result, they offer a 30-year guarantee (80% production after 30 years).

What this means is that if a 300-watt module has a –0/+5% power tolerance rating, in 10 years, most companies guarantee their modules will produce 90% (at least 270 watts) under STC in 10 years and 80% (240 watts) in 25 years under STC. SunPower guarantees that their modules will produce 80% after 30 years.

Another parameter that is listed on module spec sheets and the labels affixed to the back of modules is the *maximum system voltage.* It is the highest voltage to which an array can be wired when using that module. In residential systems, the National Electrical Code (NEC) limits maximum system voltage to 600. In commercial systems, the limit is 1,000 volts. This provision of the NEC allows installers to connect more modules in series to produce a higher-voltage system. The higher the voltage, the more efficient a system operates. (High-voltage electricity reduces line loss.)

Durability and Fire Resistance

One of the most common questions I'm asked about PV modules is: "How long will they last?" Another question is: "Will they hold up to hail?"

As noted in the sidebar on page 50, modules come with a 25- to 30-year production warranty. That's testimony to their durability and reliability. There's not a single product that I can think of with a better warranty.

Modules hold up to hail, too. In fact, all modules are designed to withstand a 25 mm (just less than 1 inch) diameter hail stone striking them at 51 miles per hour (23 mps). Faced with increasingly more violent storms, manufacturers are developing even more demanding tests with much larger hail stones striking at higher rates of speed. How prevalent is hail damage?

The answer to that question depends on where you live. In the Midwest, hail storms do occur, but they're not as common, say, as in eastern Colorado. In 2013, I inspected over 10,000 modules in 100 PV installations in western Missouri and found only a handful of modules on one array damaged by hail—and these modules were mounted flat on a roof, making them more vulnerable. The impact craters looked like they were caused by softball-sized hail.

Modules are also tested for fire resistance. This rating system has recently been improved to take into account differences in module construction—for example, whether a module is encased in glass (glass front and back) or glass in the front and plastic in back. It also takes into account the thickness of the glass. Fire ratings also factor in the type of roof and the type of rack the modules are mounted on, a topic beyond the scope of this book.

Advancements in PV: What's New and What's on the Horizon?

Although PVs have come a long way in the last few decades, researchers continue to advance the technology. They are constantly looking for ways to improve the efficiency and reduce the cost of solar electric technologies to make them more affordable to customers and profitable for businesses. This effort has led to some rather promising developments in recent years. Some of these technologies are already available; others are not ready for prime time but could be manufactured commercially in the not-too-distant future. Researchers have also developed novel applications of solar electricity like building-integrated PV (BIPV), solar roof tiles, and ways to modify arrays to increase their output. Let's start with BIPVs.

Building-Integrated PVs

BIPV incorporates photovoltaic material (and therefore, power-generating capacity) into the components of a building, for example, the roof, windows, skylights, and exterior walls. Many BIPV products incorporate some form of thin-film solar material.

On commercial buildings with flat roofs, for example, you can install a flexible roofing membrane coated with thin-film silicon. For pitched roofs, you can install solar roof tiles—special roof tiles that contain solar cells. They blend in quite nicely, but they are not yet widely used.

On standing seam metal roofs, like the one shown in Figure 3.9a, you can install a flexible thin-film material known as a *PV laminate,* or *PVL.* PV laminate (Figure 3.9b) is made by spraying thin-film material on a flexible backing such as a thin layer of aluminum or steel. It is then coated with UV-stabilized polymers (plastic). The product is manufactured in rolls. When it is time to apply PVL, a backing sheet is peeled off the laminate, revealing a sticky surface that adheres the PVL to the metal roofing.

PVL can be used for parking structures and the metal roofs of barns, homes, and other buildings. It is primarily installed on brand-new standing seam metal roofs.

As noted earlier, glass in skylights and windows can also be coated with a thin-film solar material that generates electricity. Bear in mind that glass is typically mounted vertically. Also remember that north-facing glass will receive no direct sunlight for half the year. East- and west-facing glass will be in shade half of each day. Only south-facing glass will be sufficiently illuminated. If that glass is protected by overhangs in the summer, however, solar gain will be reduced. Solar gain in south-facing glass will also be reduced throughout the summer because of the steeper altitude angle of the summer Sun. This minimizes solar gain, as discussed in Chapter 2.

Solar paints have received a significant amount of press in recent years. These are paints laced with photovoltaic material. While a cool idea, they would suffer the same problems as solar glass.

Solar modules can themselves be mounted in specially designed panels in the exterior walls of buildings—even skyscrapers—although energy production will be limited for the reasons just discussed.

As noted earlier in the chapter, some manufacturers sandwich crystalline silicon cells in glass panels that are used for the roofs of bus stop shelters or as canopies. These modules and conventional (solid-back) PV

Figure 3.9. (a) *PV laminate is applied directly to a standing seam metal roof as shown here.* Credit: National Renewable Energy Laboratory. (b) *Solar Used for Standing Seam Metal Roof. This photo shows solar laminate, or PVL, from Uni-Solar made from amorphous silicon. This product can be applied on new or existing standing seam metal roofs.* Credit: Energy Conversion Devices/Uni-Solar.

modules can be used to make awnings, carports, parking structures, patio roofs, and backyard shade structures. To me, this is the most promising BIPV. If designed correctly, structures such as these can contribute a significant amount of electricity to a home or business.

BIPVs offer several advantages over more conventional PV technologies. One of the most important advantages is that they perform multiple functions. For instance, solar deck or patio roofs, like the one shown in Figure 3.10, provide shade and lower cooling costs. They do all this while generating electricity. Such applications reduce resource depletion and construction costs.

Figure 3.10. *PV Awning. Solar awnings, like the one shown here, provide shade for windows and walls, eliminating the need for conventional awnings. Like other forms of building-integrated PVs, they perform more than one function. Awnings shade windows and walls and thus cool buildings in the summer and produce electricity year round.* Credit: Lighthouse Solar, Boulder, CO.

Another advantage of BIPVs is that they tend to blend in better than conventional PV modules. Before you embrace a BIPV technology, however, think seriously about its orientation to the Sun throughout the year. If it can't be oriented properly, its output will suffer.

Bifacial Solar Cells and Modules

Another interesting technology is the *bifacial solar cell.* Bifacial solar cells consist of a monocrystalline silicon wafer sandwiched between two ultra-thin amorphous silicon layers. Because of this, these modules can capture sunlight falling on both sides of their cells. (That's the reason they're called *bifacial* modules.)

In a bifacial module, the front side of the module generates electricity from direct and diffuse solar radiation (Figure 3.11). Diffuse solar radiation is the light that reaches the module on overcast days. The

Figure 3.11. (a) *Bifacial modules like those installed at the Midwest Renewable Energy Association harvest solar energy off both sides of the modules, although the front side shown here is by far the most productive source of electricity.* (b) *Sunlight reflecting off the light-colored roofing illuminates the back of the array, boosting its output.*
Credit: Dan Chiras.

backside generates electricity from diffuse light—that is, light bouncing off surrounding surfaces. Under the right conditions, bifacial modules can produce more power than conventional modules. In fact, their efficiency is in the range of 20.5% to 22.9%—although output is highly dependent on how they are mounted. If they're mounted on a dark-colored roof, for instance, don't expect a very significant increase in production.

When considering this or any other improved design, always be sure to compare the new type of solar module to conventional modules on the basis of the cost per unit of electricity they produce. (Electricity is measured in kilowatt-hours, or kWh.) What you will often find is that the higher cost of higher-efficiency modules does not always make economic sense. In other words, higher efficiency doesn't always pay for itself. Lower-efficiency options are often more economical. As a result, it makes more sense to add a few more conventional modules to boost the output of a PV array than to install more costly higher-efficiency modules.

Third-Generation Solar Cells

Researchers are working on numerous promising materials for solar cells. I'll focus on three of the most promising: dye-sensitized solar cells, organic solar cells, and quantum dot solar cells.

DYE-SENSITIZED SOLAR CELLS

The *dye-sensitized solar cell* (DSSC) is a type of thin-film PV. As shown in Figure 3.12, DSSCs contain particles of titanium dioxide (TiO_2) coated by a light-absorbing dye. Sunlight enters the cell through a transparent top layer, striking the dye molecules. The incoming solar radiation excites electrons in the dye molecules. These electrons are transferred directly to titanium dioxide molecules. The electrons then move to the clear surface layer, which doubles as the contact.

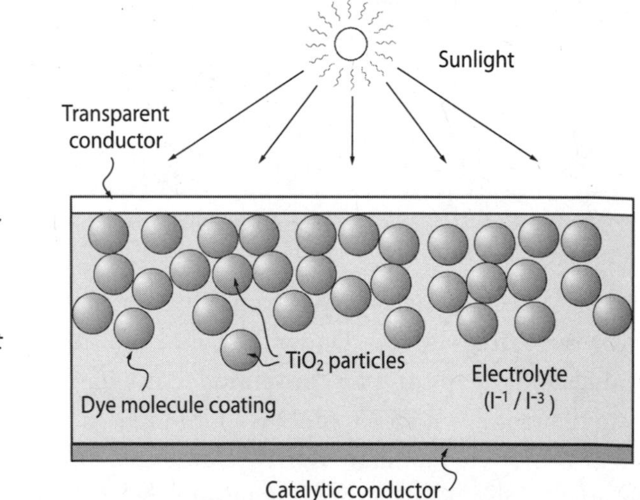

Figure 3.12. *Dye-Sensitized Solar Cell. This drawing shows the anatomy of one of the newest and most promising solar cells under development.* Credit: Anil Rao.

This technology uses low-cost materials and can be used to manufacture long, semi-flexible and semi-transparent rolls. As a result, cells can be assembled using much less energy with much simpler and less expensive equipment than that required to manufacture conventional monocrystalline and polycrystalline PVs.

On the downside, manufacturers have had a tough time eliminating a number of expensive materials needed to manufacture this product (platinum and ruthenium). The liquid electrolyte is also subject to freezing, making it quite challenging to design a cell suitable for cold regions. Another problem is their low conversion efficiency. DSSCs are currently about 11% efficient, although efficiencies of 15% have been attained in the laboratory.

Since the previous edition of this book, dye-sensitized solar cells have made their way into the market as power sources on sports bags that charge portable electronic devices. Some believe that commercial production is just around the corner, but, as Yogi Berra said, "It's tough to make predictions, especially about the future."

ORGANIC SOLAR CELLS

Another technology that holds promise is the *organic solar cell.* Organic solar cells consist of ultrathin plastic films containing organic (carbon-based) materials. These materials are capable of emitting electrons when illuminated by sunlight and thus produce DC electricity. (Although they do it in a very different manner.)

The simplest organic cells consist of a single layer of photo-excitable and conductive organic material sandwiched between two conductors (labeled "electrodes" in Figure 3.13). Understanding how the cells

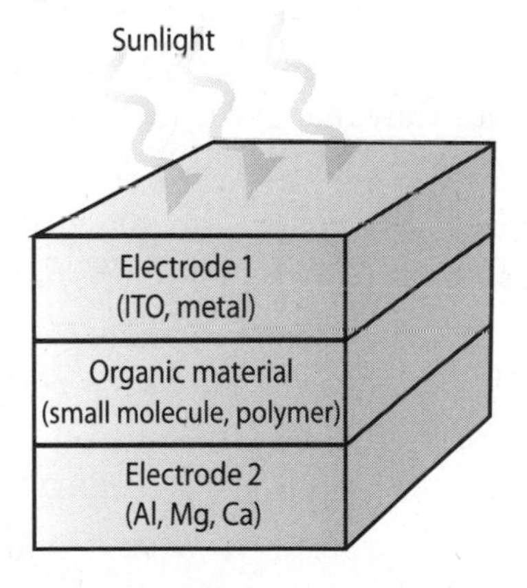

Figure 3.13. *Single-layer Organic Solar Cell. The simplest (and least efficient) organic solar cell consists of a single layer of reactive material.* Credit: Forrest Chiras.

generate electricity requires a Ph.D. in physics. A master's degree in organic chemistry might also help. If you're interested, you can find excellent descriptions on the internet.

The organic molecules used to make organic PVs can be produced in mass quantities and relatively inexpensively. Moreover, the molecules currently being tested are also capable of absorbing a large amount of solar energy per unit of material, potentially providing higher efficiencies. In addition, slight chemical modifications of these molecules can result in a variety of compounds that absorb different wavelengths of solar radiation, which could also improve the efficiency of multi-junction cells. In fact, most of the research is focused on creating cells that have multiple layers of different materials. Finally, because they use so little material and the materials are inexpensive, organic cells could eventually be produced at a very low cost. (That remains to be seen, however.)

Unfortunately, to date, most organic cells produced in the laboratory are rather inefficient. The highest efficiencies range from 6.5% to 11.6%. (Expect that to change.) Organic solar cells have the additional drawback of breaking down easily in sunlight, although efforts are underway to address this issue.

Quantum Dot Solar Cells

Another interesting development is the *quantum dot solar cell.* This technology draws on the relatively new and exciting branch of science called *nanochemistry.* Like other third-generation solar cells, this technology relies on materials that absorb a variety of wavelengths of light. It's therefore ideal for multi-junction solar cells.

Quantum dots (QDs) are tiny semiconducting particles. Their ability to absorb different wavelengths of light can be changed by altering their size. Different size particles can be produced without any changes in the material.

Single-junction QD cells are made from lead sulfide (PbS). They can capture long-wave heat radiation to generate electricity, something not possible with traditional solar cells. (Remember that 50% of the Sun's output is infrared radiation.) This characteristic could dramatically increase the output of a solar cell. Some scientists believe quantum dot solar cells could eventually achieve efficiencies of nearly 70%. So far, however, efficiencies in this new technology are fairly low—between 7% and 8.7%. (Expect those numbers to change.)

Quantum dot solar cells should be inexpensive to produce. Materials are cheap, and they are made with relatively simple and inexpensive equipment. The dots are "sputtered" onto a substrate. In large-scale production facilities, they could be sprayed on, like ink.

Conclusion: Should You Wait for the Latest, Greatest New Technology?

Knowing that newer high-efficiency PV technologies are in the offing, does it make sense to delay your installation until the new, PV technologies hit the market?

The answer is no.

First, the new technologies are either not commercially available or they won't be for some time. Don't sit around waiting for them. When they do become available, they may be quite costly. The electricity they produce could cost a lot more than electricity generated by mass-produced, albeit lower efficiency, solar modules.

Be content with the high-efficiency monocrystalline and polycrystalline modules on the market today. They're capable of producing electricity inexpensively for a long, long time. Go for it!

CHAPTER 4

Does Solar Electricity Make Sense?

Sizing a System and Evaluating Its Cost and Economic Benefits

Whether you are going to install a solar electric system yourself or hire a professional, one of the first things you'll need to determine is the size of the system. The size of the system is dictated primarily by your electrical demands and the percentage of your electricity you would like to produce via a solar electric system.

Once you know the system size, you'll most likely want to know how much it is going to cost. Will your investment be worth it? Would you be better off continuing to buy utility power?

Let's say you do find that a solar electric system could meet your needs. What can you do if you're not able to come up with the money? Don't despair, there are alternatives like community solar and leasing programs that can get you into a solar electric system at a lower cost or no cost at all.

In this chapter, I'll walk you through these steps. We'll begin with assessing your electrical demand.

Assessing Electrical Demand

To size a PV system, you first need to know how much electricity you consume. How you determine the amount of electricity you consume depends on whether you are planning to power an existing home or a new one—either a home you are buying or one you are building. Let's begin with existing structures.

Assessing Electrical Demand in an Existing Home

Assessing electrical consumption in an existing home is as simple as tallying monthly consumption from one's electric bills. I recommend going back at least two years—preferably three. If you have recycled your bills, call the utility or log on to your account to find the information. What you are trying to determine is the number of kilowatt-hours of electricity you consume each year.

If you're like most people, annual electrical demand in kilowatt-hours is all you need to size your system. (I'll show you how, shortly.) If you are thinking about severing your ties with the utility—that is, taking your house off the grid—you will need to determine monthly averages. Off-grid systems are sized to meet demands during the times of highest consumption.

After you have calculated annual energy consumption (in the case of a grid-connected PV system) or monthly averages (in the case of an off-grid system), take a few moments to look for trends in energy use. Is energy consumption on the rise? Or is it staying constant, or declining? If you notice an increase in electrical use in recent years, perhaps because you added air conditioning, discard earlier energy data and recalculate the averages based on the most recent bills. Early data will lower the average, and it won't represent current consumption. You could end up undersizing your PV system.

If, on the other hand, electrical energy consumption has declined because you replaced your inefficient furnace and refrigerator with newer, more frugal models, earlier data will artificially inflate electrical demand. You'll need less electricity than the averages indicate and a smaller system.

Assessing Electrical Demand in a New Home

Determining electrical demand in a brand-new home—one that's just been built or one that is about to be built—is much more challenging.

If you're purchasing a new home and thinking about installing solar, one of the best ways to determine electrical demand is to live in it for a year or two to gather data. Or, you can obtain energy data from the previous owner—if they've saved their utility bills and are willing to share this information with you. If they don't have their bills, they may be willing to contact the local utility company to request a summary of their electrical consumption over the past two to three years. Remember, however, a house does not consume electricity, its occupants do, and we all use energy differently. If the previous homeowners used energy wastefully, their energy consumption data may be of little value to you if you are an energy miser.

Another method to estimate electrical consumption is to base it on the electrical consumption of your existing home or business. For instance, suppose you are building a brand-new home that's the same size as your current home. If the new home will have the same amenities and the same number of occupants, electrical consumption could be similar to your existing home.

If, however, you are building a more energy-efficient home and are installing much more energy-efficient lighting and appliances and incorporating passive solar heating and cooling (all of which I highly recommend), electrical consumption could easily be 50%, perhaps 75%, lower than in your current home. If that's the case, adjust your electrical demand to reflect your new, more efficient lifestyle.

A more scientific way to estimate electrical consumption is to perform a load analysis. A load analysis is an estimate of electric consumption based on the number of electronic devices in a home, their average daily use, and energy consumption. It is a lot more difficult to calculate than you would expect.

To perform a load analysis, you or an installer begin by listing all the appliances, lights, and electronic devices that will be used in the new home. Rather than list every light bulb separately, however, you may want to lump them together by room. I like to work with clients one room at a time, using a spreadsheet I've prepared for each project. Or, you can use a worksheet like the one shown in Table 4.1. Similar worksheets can be

Table 4.1. Load Analysis

Individual Loads	Quantity x	Volts x	Amps =	Watts AC	Watts DC x	Use (hrs/day) x	Use (day/wk) ÷	7 days +	Watts Hours AC	Watts Hours DC
								7		
								7		
								7		
								7		
								7		
								7		
								7		
								7		
								7		
								7		

AC Total Connected Watts: ______________ AC Average Daily Load: ______________

DC Total Connected Watts: ______________ DC Average Daily Load: ______________

Note: The column labeled DC is for those folks who want to install an off-grid system and power some or all of their loads with DC electricity. FYI, DC wattages are the product of DC amps × DC volts. A device that uses 1 amp at 48 volts requires 48 watts.

found online at Northern Arizona Wind and Sun's website (solar-electric.com) and other websites.

Once you've prepared a complete list of all the devices that consume electricity, you need to determine how much electricity each one uses. The amount of electricity consumed by an appliance, lamp, or electronic device can be determined by consulting a chart like the one in Table 4.2. Charts such as this list typical wattages for a wide range of electrical devices. Detailed listings are available online.

Table 4.2. Average Electrical Consumption of Common Appliances

Appliance	Watts
General household	
Air conditioner (1 ton)	1500
Alarm/Security system	3
Blow dryer	1000
Ceiling fan	10–50
Central vacuum	750
Clock radio	5
Clothes washer	1450
Dryer (gas)	300
Electric blanket	200
Electric clock	4
Furnace fan	500
Garage door opener	350
Heater (portable)	1500
Iron (electric)	1500
Radio/phone transmit	40–150
Sewing machine	100
Table fan	10–25
Waterpik	100
Refrigeration	
Refrigerator/freezer 22 ft^3 (14 hrs/day)	540
Refrigerator/freezer 16 ft^3 (13 hrs/day)	475
Sun Frost refrigerator 16 ft^3 (7 hrs/day)	112
Vestfrost refrigerator/freezer 10.5 ft^3	60
Standard freezer 14 ft^3 (15 hrs/day)	440
Sun Frost freezer 19 ft^3 (10 hrs/day)	112
Kitchen appliances	
Blender	350
Can opener (electric)	100
Coffee grinder	100
Coffee pot (electric)	1200
Dishwasher	1500
Exhaust fans (3)	144
Food dehydrator	600
Food processor	400
Microwave (.5 ft^3)	750
Microwave (.8 to 1.5 ft^3)	1400
Mixer	120
Popcorn popper	250
Range (large burner)	2100
Range (small burner)	1250
Trash compactor	1500
Waffle iron	1200
Lighting	
Incandescent (100 watt)	100

Incandescent light (60 watt)	60	Computer (laptop)	20–50
Compact fluorescent (60 watt equivalent)	16	Electric player piano	30
		Radio telephone	10
Incandescent (40 watt)	40	Satellite system (12 ft dish)	45
Compact fluorescent (40 watt equivalent)	11	Stereo (avg. volume)	15
		TV (31.5-inch color)	22–30
		TV (40-inch color)	28–37
Water Pumping		TV (48-inch color)	35–55
AC Jet pump (¼ hp) 165 gal per day, 20 ft. well	500	VCR	40
DC pump for house pressure system (1–2 hrs/day)	60	**Tools**	
		Band saw (14″)	1100
DC submersible pump (6 hours/day)	50	Chain saw (12″)	1100
		Circular saw (7¼″)	900
		Disc sander (9″)	1200
Entertainment		Drill (¼″)	250
CB radio	10	Drill (½″)	750
CD player	35	Drill (1″)	1000
Cellular telephone	24	Electric mower	1500
Computer printer	100	Hedge trimmer	450
Computer (desktop)	80–150	Weed eater	500

For more accurate data, I recommend that you check out the nameplates on the appliances and electronic devices you'll be installing. The nameplate is a sticker or metal plate listing a unit's electrical consumption. The measure you are looking for is watts. Wattage is a measure of instantaneous power consumption. Manufacturers sometimes list amps and volts. If the nameplate lists amps, but not watts, simply multiply the amps × volts to calculate the wattage (amps × volts = watts). Most electronic devices are plugged into 120-volt outlets.

Multiplying amps by volts yields the wattage of many electronic devices, except those with electric motors known as *induction motors*. These include fans, washing machines, clothes dryers, dishwashers, pumps, and furnace blowers. In such cases, multiplying amps by volts significantly overestimates the wattage. For these devices, calculate wattage by multiplying amps by volts, then multiply wattage by 0.6 to obtain a closer estimate of true wattage.

The nameplate wattage of other devices, such as televisions, stereos, and power tools, is also typically higher than the wattage the appliance will draw when in normal operation. That's because, the wattage listed on an appliance nameplate is the power the device draws at maximum load. Most devices rarely operate at maximum load. Reducing the nameplate wattage by 25% is a reasonable adjustment for such devices, although typical operating wattage may be even lower.

A more accurate way to determine the wattage of an electronic device is to measure it directly using a meter like those shown in Figure 4.1. To use it, plug the meter into an electrical outlet and then plug the appliance into the device. A digital readout indicates the instantaneous power (watts). Much easier!

When measuring a device that cycles on and off, such as a refrigerator, leave the watt-hour meter connected for a week or two. The meter will record the total energy used during this period in kilowatt-hours; it will also tell you how many hours you have been recording data. You can then calculate the number of kilowatt-hours the device consumes in a 24-hour day.

Power consumption can also be determined by checking spec sheets for various electronic devices. You can even go online to find them.

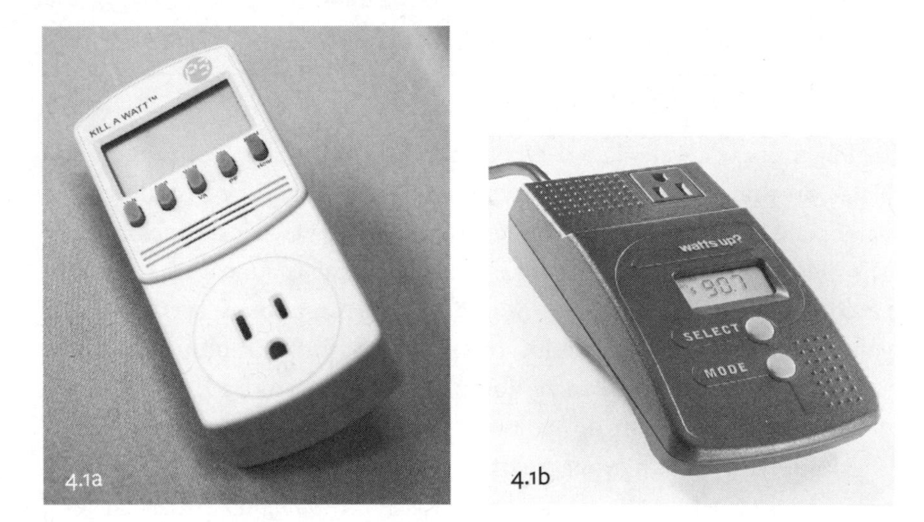

Figure 4.1. (a) *The Kill A Watt and* (b) *Watts Up? meters can be used to measure wattage of household appliances and electronic devices. Of the two, the Watts Up? is the more sensitive and allows for measurement of tiny phantom loads as well.* Credit: (a) Dan Chiras, (b) Electronic Educational Devices.

After you have determined the wattage of each electrical device, you must estimate the number of hours each one is used on an average day and how many days each device is used during a typical week. From this information, you calculate the weekly energy consumption of all devices in your home or business. You then divide this number by seven to determine your average daily consumption in watt-hours or kilowatt-hours. You can use this number to estimate annual energy consumption.

Load analysis is an imprecise art that is fraught with problems. One shortcoming of this technique is that it is often difficult for individuals to estimate how long each appliance runs on a typical day. For example, it is difficult to estimate how many minutes you operate your toaster and blender each day. Is it three minutes, five minutes, or ten? How long does the refrigerator run? What about your well pump? Or sump pump? Moreover, most people tend to underestimate TV time. And what about the kids? How many hours of television do they watch each day? How many hours a day do they spend on their computers or cell phones?

Another problem with this process is that run times vary by season. Electrical lights, for example, are used much more in the winter, when days are shorter, than in the summer. Furnace blowers operate a lot during the winter, but not at all or very infrequently the rest of the year. So, what's the average daily run time for the furnace blower?

Another problem with this approach is that many electronic devices draw power when they're off. Such devices are known as a *phantom loads.* They include television sets, VCRs, satellite receivers, cable boxes, and power cubes like those we use to charge laptop computers and other portable electronic devices. How do you locate phantom loads?

Basically, any electronic device that has an LED light that shines all the time is a phantom load. All remotely controlled devices with an instant-on feature also consume power when they are off, and are therefore phantom loads. Devices like coffee makers and microwaves with LED clocks are phantom loads.

Phantom loads are often quite small, 1 watt or less, though some can be quite significant, in the range of 10 to 20 watts. No matter what size they are, when added together, they quickly add up. Also, because they're drawing power 24 hours a day, your combined phantom loads can easily account for 5% of your annual electrical consumption. Bottom line: be sure to include phantom loads when determining your total electrical consumption. To locate phantom loads, plug them into a plug-in watt-hour meter. That will tell you how many watts each one requires.

Because of the various problems I've discussed, homeowners and installers often grossly underestimate their electrical consumption. If you estimate your electrical demand this way, it's a good idea to bump up your estimate by at least 25% just to be sure.

Once you've calculated daily electrical energy use, it's time to size a solar electric system, right?

Actually, no.

Before you size a PV system, it's a good idea to look for ways to reduce electrical use through energy-efficiency and conservation measures. Don't forget: the lower the energy demand, the smaller the solar electric system you'll need. The smaller the system, the less you'll spend. In addition, because smaller systems require fewer resources to manufacture, you're helping reduce your environmental impact, including your carbon footprint.

Remember, too, that money spent on energy-efficiency measures may have a better return on investment than money spent on a PV system. An

Figure 4.2. *Phantom Load. This superefficient, 50-inch Samsung flat-screen television uses only 60 to 130 watts when operating. Because of the instant-on feature, however, it continues to draw 8 watts when turned off. The satellite receiver attached to the TV uses 26 watts when operating and 24 watts when turned off. To save energy, homeowners can plug electronics like this into power strips that are shut off when the unit is not on. Or, a homeowner can install an outlet controlled by a switch; however, these must be installed during construction.* Credit: Dan Chiras.

energy-efficient refrigerator, for instance, that saves you 1,000 kWh a year might cost $800 to $900. A PV array that would generate 1,000 kWh a year will cost you $2,500, maybe a bit more. How do you determine the most cost-effective measures to reduce electrical consumption?

The best way to devise an energy-efficiency strategy is to hire an experienced home energy auditor. He or she will perform a thorough energy analysis of your home and make recommendations on ways you can reduce your electric bill. A good auditor will also prioritize energy-saving measures—that is, select the options that are the most cost effective. If you are a do-it-yourselfer, you can perform your own home energy audit, but that's a sizeable task. For guidance, check out my book, *Green Home Improvement.*

Some solar installers may provide guidance on energy savings as well. Unfortunately, in my experience, most installers are not well versed in efficiency measures. They are there to sell you a solar system—and the bigger the better. Some inexperienced or unscrupulous installers will also recommend installing PV systems in less-than-optimum sites—for example, on east- or west-facing roofs (when south-facing roofs are small, nonexistent, or are shaded). Read this book carefully so you don't get led astray.

Conservation and Efficiency First!

Before you invest in a solar electric system, I highly recommend making your home—and your family—as energy efficient as possible. Even if you're an energy miser, you may be able to make significant cuts in energy use.

Waste can be slashed many ways. Interestingly, though, the ideas that first come to mind for most homeowners tend to be the most costly: new windows and energy-efficient washing machines, dishwashers, furnaces, or air conditioners (Figure 4.3). While vital to creating a more energy-efficient way of life or business, they're the highest fruit on the energy-efficiency tree—and the most expensive.

Before you spend a ton of money on windows or new appliances, I strongly recommend that you start with the lowest-hanging fruit. These are the simplest and cheapest improvements, and will yield the greatest energy savings at the lowest cost.

Huge savings can be achieved by changes in behavior, too. You've heard the list a million times: Turn off lights, stereos, computers, and TVs when

Figure 4.3. *Energy-efficient Washing Machine. Spending a little extra for an energy-efficient front-loading washing machine like this (bottom unit) will reduce the size of your solar system. To dry clothes, though, consider using a solar clothes dryer (commonly called a clothesline).* Credit: Frigidaire.

not in use. Turn your thermostat down a few degrees in the winter. Wear sweaters and insulated underwear. Turn the thermostat up in the summer and run ceiling fans. Open windows to cool a home naturally, especially at night, in the spring and summer, and then close windows in the morning to keep heat out during the day. Draw the shades or blinds on the south, east, and west side of your house in the summer to reduce cooling costs. All these changes cost nothing, except a little of your time, but they can reap enormous savings—not just in your monthly energy bill, but also in the cost of a PV system.

Other low-hanging, high-yield fruit include tightening up our homes and workplaces, that is, making them more airtight. Weather-stripping around doors and caulking leaks in the building envelope can reap huge benefits. Once a building is better sealed, beef up the insulation. Install insulated window shades. Insulate attics, walls, and floors above crawl spaces or garages. For advice on insulation, be sure to hire an energy consultant. If they're good, they can help you figure out the best ways to seal and insulate your home.

Once you've made these changes, it's time to examine bigger-ticket items. For example, your refrigerator. In many homes, refrigerators are responsible for a staggering 25% of total electrical consumption. If your refrigerator is old and in need of replacement, unplug the energy hog, recycle it, and replace it with a more energy-efficient model. Thanks to dramatic improvements in design, refrigerators on the market today use significantly less energy than refrigerators manufactured 20 years ago. (And please, do *not* plug in the old one in the basement or out in the garage. That would defeat the entire enterprise!)

Electrical energy use can also be reduced by replacing energy-inefficient electrical devices with newer, more efficient Energy Star models (Figure 4.4). Before you go shopping, log on to the Energy Star website. Click on the appliance or electronic device you're interested in. Look for models that meet your criteria. *Consumer Reports* also has an excellent website that lists energy-efficient appliances. Their site also rates appliances on reliability. Canadian readers can log on to oee.nrcan.gc.ca for a list of international Energy Star appliances.

When shopping for appliances both online and in stores, you can compare the energy efficiency of refrigerators, freezers, and other devices by checking out the yellow Energy Guide on or inside the device. It will tell you how much electricity a particular appliance will typically use in a year's time and how the model you are looking at compares to other models in that category. Figure 4.5 shows an example of an Energy Guide. In Canada, the same yellow tags are also posted on appliances, but it's called an EnerGuide.

Figure 4.4. *Energy Star Label. When shopping for electronic devices such as computers, stereo system components, and television sets, be sure to look for the Energy Star label. This label indicates that the product is one of the most energy efficient in its category.* Credit: EPA.

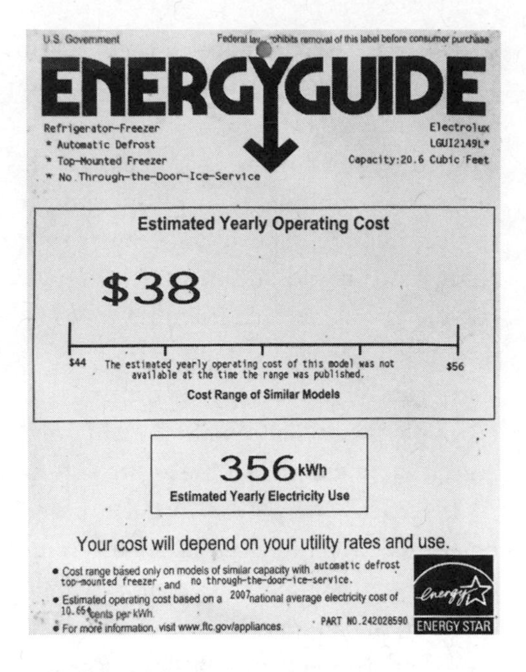

Figure 4.5. *Energy Guide Tag. When shopping for appliances such as refrigerators and freezers, be sure to look for the Energy Guide, like the one shown here. These guides indicate how much electricity an appliance will use in one year and how the appliance you are looking at compares to others in its product category. This one is for the refrigerator/freezer that we purchased for our home; it uses the least energy of all the Energy Star-rated refrigerators in its category.* Credit: Dan Chiras.

You can also trim some of the fat from your energy diet by installing more energy-efficient lighting, such as compact fluorescent lights or, better yet, longer-lasting and slightly more efficient LED lights.

Although efficiency has been the mantra of energy advocates for many years, don't discount its importance just because the advice has grown a bit threadbare. As it turns out, very few people have heeded the persistent calls for energy efficiency. And, many who have made changes have not fully tapped the potential savings.

Readers interested in learning more about making their homes energy efficient may want to read the chapters on energy conservation in one of my books, *Green Home Improvement* or *The Homeowner's Guide to Renewable Energy.*

Sizing a Solar Electric System

Once you've estimated your electrical demand and devised and implemented a strategy to use energy more efficiently, it is time to determine your system's size. This is a step usually performed by professional installers. In the following material, I'll describe how to size all three types of solar systems: (1) grid-connected, (2) grid-connected with battery backup, and (3) off-grid.

Sizing a Grid-Connected System

Most systems installed these days are grid-connected. They do not require batteries. These systems produce electricity rain or shine. The electricity is consumed in the home. If there's a surplus, it is backfed onto the electrical grid. The utility keeps track of the surpluses fed onto the grid. At night, the utility supplies electricity to the home, but there's no charge for it if there's been a surplus fed onto the grid.

Sizing a grid-connected system is the easiest of all. To meet 100% of your needs, determine how much electricity you consume in a year, then divide that by 365 to find the average daily electrical demand (in kilowatt-hours).

This number is then divided by the average peak sun hours per day for your area. As noted in Chapter 2, peak sun hours can be determined from solar maps and from various websites and tables.

As an example, let's suppose that you and your family consume 12,000 kWh of electricity per year (or 1,000 kWh per month). This is about 34 kWh per day. Next, divide 34 kWh per day by peak sun hours.

Let's suppose you live in Lexington, Kentucky, where the average peak sun hours per day is 4.5. Dividing 34 by 4.5 gives you the array size (capacity): 7.6 kilowatts. But don't run out and order a system quite yet.

To calculate the size, you need to increase it by 22%. This accounts for a number of factors that lower the production of a PV system in the field such as high temperatures, dust on modules, losses due to voltage drop as electricity flows through wires, and inefficiencies of various components (inverters, controllers, etc.).

To calculate the array size using the 22% adjustment factor, you simply divide 7.6 kW by 0.78. In our example, then, we'd need to install a 22% larger PV array, one rated at 9.74 kW. A 9.74 kW array would require thirty four 285-watt modules.

If your site is shaded, the system may need to be larger because shading lowers the output of a PV system. To determine the amount of shading on a PV system, professional solar site assessors and solar installers often use a Sun path analysis tool like the Solar Pathfinder shown in Figure 4.6. The Solar Pathfinder and similar devices determine the percentage of solar radiation blocked by local features in the landscape such as trees, hills, and buildings. If the device indicates 10% shading, the array would need to be 10% larger. Alternatively, trees or branches could be trimmed to eliminate shading. If shading is severe, an alternative location would be required. For optimum performance, PV systems generally require a site where an array is unshaded 90% to 95% of the time.

Figure 4.6. *Solar Pathfinder. This device allows solar site assessors and installers to assess shading at potential sites for PV systems, helping them select the best possible site to install an array.* Credit: Shawn Schreiner.

Figure 4.7. *Solar Pathfinder Dome Showing Shading. The dome of the Solar Pathfinder (not shown here) reflects all obstructions that will shade a solar array. A photograph of the dome is then entered into a computer program that allows the operator to trace the shading. The computer then calculates the amount of shading that occurs in each month and determines how much electricity an array could produce with that amount of shade.* Credit: Solar Pathfinder.

Sizing an Off-Grid System

Off-grid solar electric systems are sized according to the month of the year with the highest demand. In the Canada and northern-tier states in the US, that month is typically January—it's the coldest month and the month that offers the least sunshine. In southern states, peak demand often occurs in the summer when air conditioning is working overtime to keep us cool.

After determining the month with the highest demand, calculate the average daily demand for that month. Then look up the average peak sun hours for the month. (This is available on various websites.) Next divide the average daily consumption by the average peak sun hours to determine the size of the array. Don't forget to adjust for efficiency and shading. For battery-based systems, a 30% efficiency factor is a good idea. In other words, you should increase the array size by 30%.

As an example, let's suppose that a superefficient off-grid home requires 200 kWh of electricity during the month of January. That's 6.5 kWh per day. If the average daily peak sun for January is 2.9, you divide 6.5 kWh per day by 2.9. The result is a 2.5 kW system. Now adjust by 30%. To do so, divide 2.5 by 0.70. The result is 3.6 kW. That's the size of the solar electric system you'd need to install.

Once you know the size of your array, you must size the battery bank. Although I'll discuss battery bank sizing in detail in Chapter 7, it's important to note that battery banks are sized to provide sufficient electricity to meet a family's or business' needs during cloudy periods. We call these "battery days." Most battery banks are based on a three-to-five-day reserve so they'll provide enough electricity for three to five days of cloudy weather. In sunny Colorado, New Mexico, or Utah, I'd size the battery bank for three battery days. In cloudier Ohio, Indiana, or Illinois, I'd size a system for five battery days.

Sizing a Grid-Connected System with Battery Backup

PV arrays in grid-connected systems with battery backup are sized much like grid-connected systems. That is, the size of the array is based on average daily demand. In these systems, however, you'll also need to size a battery bank.

Battery banks are sized according to the electrical requirements during a power outage. However, to save money, most homeowners opt only to power critical loads during these periods. Critical loads are devices you must have such as pumps and fans of heating and cooling systems, well pumps, sump pumps, refrigerators, freezers, and a few lights in critical areas, such as kitchens. Because these loads require less energy than the entire home, battery banks in these systems are typically much smaller than in off-grid systems. In fact, they are about one-third of the size.

Does a Solar Electric System Make Economic Sense?

At least three options are available to analyze the economic cost and benefits of a solar electric system: (1) a comparison of the cost of electricity from the solar electric system vs. conventional power; (2) simple return on investment; and (3) a more sophisticated economic analysis tool known as *discounting.*

Cost of Electricity Comparison

One of the simplest ways of analyzing the economic performance of a solar system is to compare the cost of electricity produced by a PV system to the cost of electricity from your utility.

Once you know the size of your system, how much it will produce each year, and the cost to install it, you first calculate the system's output over a 30-year period—the expected life of a PV system. Next, you calculate the cost per kilowatt-hour. You do this by dividing the cost of the PV system by the total output in kWh over 30 years.

Suppose you live in sunny western Colorado and are interested in installing a grid-connected solar electric system that will meet 100% of your electric needs. Your superefficient home requires, on average, 500 kWh of electricity per month, or 6,000 kWh per year. That's 16.4 kWh per day. Peak sun hours in your area is 6. To size the system, divide the electrical demand (16.4 kWh per day) by the peak sun hours. The result is 2.7 kW. Adjusting for 78% efficiency, the system should be 3.46 kWh. Let's round up to 3.5 kW. Let's assume that the system is not shaded at all during the year.

Your local solar installer says she can install the system for $3.14 per watt (the national average in January 2019). The system will therefore cost $10,990 (3,500 watts x $3.14 per watt). Next, subtract the 26% tax credit (available in 2020) from the federal government from the cost of the system. (The credit is scheduled to drop to 20% in 2021 and 10% in 2022.)

The federal tax credit is based on the cost of the system, including installation ($10,990), minus state or utility rebates (if any). In this case, let's assume there are no state or utility rebates. Twenty-six percent of $10,990 equals $2,857. Total system cost after subtracting this incentive is $8,133.

According to your calculations or the calculations provided by the solar installer, this system will produce, on average, 6,000 kWh per year. If the system lasts for 30 years, it would produce 180,000 kWh over this period.

To calculate the cost per kilowatt-hour, divide the system cost ($8,133) by the output (180,000 kWh). In this case, your electricity will cost 4.5 cents per kWh. Considering that the going rate in Colorado is currently over 11 cents per kWh, the PV system represents an excellent investment. Bear in mind, too, that utility costs will continue to increase over the next 30 years. Note, too, that even without the federal tax credit, the cost of electricity from a solar electric system would be still be low—about 6.1 cents per kilowatt-hour over its 30-year lifespan.

If you'd like to learn more about the ins and outs of this calculation, you might want to check out my book, *Power from the Sun,* now out in its second edition.

Calculating Simple Return on Investment

Another method used to determine the cost-effectiveness of a PV system is *simple return on investment* (ROI). Simple return on investment is, as its name implies, the *savings* generated by installing a PV system, expressed as a percentage.

Simple ROI is calculated by dividing the annual dollar value of the energy generated by a PV system by its cost. A solar electric system that produces 6,000 kWh of electricity per year that would cost you 11 cents per kilowatt-hour when purchased from the utility generates $660 worth of electricity each year. If the system costs $8,133, after rebates, the simple return on investment is $660 divided by $8,133 × 100 which equals 8.1%. If the utility charges 15 cents per kWh, the 6,000 kWh of electricity would be worth $900, and the simple ROI would be 11%. Both of these represent very decent rates of return. Consider what your money would be doing otherwise. In the States, the average interest on savings accounts in banks was 0.09% in 2019. The top online savings accounts offer a whopping 2.1 to 2.3% at this writing, but you might need a balance of $20,000 to receive this rate!

Weaknesses of Economic Analysis Tools

Comparing the cost of electricity and return on investment are both simple tools. However, both fail to take into account a number of economic factors. For example, both techniques fail to account for interest payments on loans that may be required to purchase a PV system. Interest payments will add to the cost of electricity produced by the system. For those who self-finance, for example, by taking money out of savings, both tools fail to take into account opportunity costs—lost income from interest-bearing accounts raided to pay for the system. These calculations also don't take into account possible maintenance, possible increases in property tax, possible increases in insurance premiums, and inverter replacement. All these costs will decrease the economic benefits of a solar electric system.

That said, these simple economic tools also leave the rising cost of electricity out of the equation. Nationwide, electric rates have increased on average about 4.4% per year over the past 35 years. In recent years, the rate of increase has been double that in some areas. These methods also don't give credit for the large increase in home values produced by the presence of a solar electric system and/or much lower utility bills.

All in all, the two sets of factors—those that raise the system costs and those that make it more profitable—very likely offset each another, so the cost of electricity comparison and the simple return on investment turn out to be fairly valuable tools for analyzing the economic sensibility of a PV system. They're infinitely better than the old standby, *payback* (also known as *simple payback*).

Payback is a term that gained popularity in the 1970s first in relation to energy-efficiency measures, then to solar systems. Payback is the number of years it takes a renewable energy system or energy-efficiency measure to pay back its cost through the savings it generates. Payback is calculated by dividing the cost of a system by the anticipated annual savings. If the $8,133 PV system we've been looking at produces 6,000 kWh per year, and grid power costs you 11 cents per kWh, the annual savings of $660 yields a payback of 12.3. years. From that point on, the system produces free electricity.

While the payback of 12.3 years seems ridiculously long, don't forget that this is the system that yielded a very respectable 8.1% return on investment, which looked pretty good. While payback is popular concept among buyers, this example shows that it has very serious drawbacks. The most important is that what appears to be a relatively long payback actually represents a pretty good ROI.

Net Present Value: Comparing Discounted Costs

For those who want a more sophisticated tool to determine whether an investment like a PV system makes sense, economists have devised an ingenious method that allows us to compare the value of a solar electric system to the cost of buying electricity from a utility for the next 30 years. It allows us to make this rather tricky comparison based on the value of today's dollar—or whatever currency your country uses. They call the value of something in today's dollar *present value.*

Unlike the previous methods, this method takes into account numerous additional economic factors besides the cost of the system. These can include maintenance costs, insurance, inflation, and the rising cost of grid power.

As all readers know, inflation decreases the value of money over time. Economists refer to this as the *time value of money.* The time value of money takes into account the fact that a dollar a year from now will be worth less than a dollar today. Economists refer to the rate at which the value of money declines as the *discount factor.*

To make life easier, this economic analysis can be performed by using a spreadsheet. The details of this method are beyond the scope of this book, but let me say just one thing: In most cases, this technique illustrates that an investment in a PV system is far better than continuing to pay your electric utility for 30 more years. Table 4.3 shows an analysis done on a solar electric system installed in Colorado. If you look at the bottom row, you will see the cost of buying electricity for 30 years. It's over $34,000. In present day dollars, that's over $20,000. The solar system with a new inverter installed after 20 years will have cost you nearly $13,000. Adjusted for the time value of money, your cost in today's dollars would be $9,555. As you can see, it is much more cost effective to go solar than to pay the utility. To learn more about this technique, you can check out my book, *Power from the Sun.*

Table 4.3. Economic Analysis of PV System

Year	Discount Factor 3.0%	Buy Utility Electricity		Proposed PV System	
		Cost 4.4%	Discounted Cost	Cost	Discounted Cost
0	1.000	$0	$0	$9,555	$10,710
1	0.971	$570	$553	$0	$0
2	0.943	$595	$561	$0	$0
3	0.915	$621	$569	$0	$0
4	0.888	$649	$576	$0	$0
5	0.863	$677	$584	$0	$0
6	0.837	$707	$592	$0	$0
7	0.813	$738	$600	$0	$0
8	0.789	$771	$608	$0	$0
9	0.766	$804	$617	$0	$0
10	0.744	$840	$625	$0	$0
11	0.722	$877	$633	$0	$0
12	0.701	$915	$642	$0	$0
13	0.681	$956	$651	$0	$0
14	0.661	$998	$660	$0	$0
15	0.642	$1,042	$669	$0	$0
16	0.623	$1,087	$678	$0	$0
17	0.605	$1,135	$687	$0	$0
18	0.587	$1,185	$696	$0	$0
19	0.570	$1,237	$706	$0	$0
20	0.554	$1,292	$715	$3,200	$1,772
21	0.538	$1,349	$725	$0	$0
22	0.522	$1,408	$735	$0	$0
23	0.507	$1,470	$745	$0	$0
24	0.492	$1,535	$755	$0	$0
25	0.478	$1,602	$765	$0	$0
26	0.464	$1,673	$776	$0	$0
27	0.450	$1,746	$786	$0	$0
28	0.437	$1,823	$797	$0	$0
29	0.424	$1,903	$808	$0	$0
30	0.412	$1,987	$819	$0	$0
Total		**$34,191**	**$20,330**	**$12,755**	**$9,555**

Alternative Financing for PV Systems

Many individuals and businesses do not have the financial wherewithal to purchase a PV system outright, even with various incentives. Fortunately, there are some alternative financing mechanisms that could make your dreams of a PV system come true: *leases, community solar,* and *co-op solar.*

Leases are a pretty painless way to get into solar, which is why they are so popular. According to recent estimates, about two-thirds of all solar electric systems are currently being installed under leasing arrangements.

Leases vary from one company to the next. Typically, companies install PV systems for their customer at no cost. Some companies may offer the homeowner an option to invest in the system in trade for lower electrical bills. Companies typically agree to repair any damage they might cause when installing a system, especially roof damage. The company, in turn, owns, monitors, insures, and maintains the system for the duration of the agreement.

Leases are long-term agreements, typically 20 years. At the end of the lease, customers can renew their agreement or enter into a new agreement, which often entitles them to newer equipment. If a customer wishes to terminate his or her agreement, the companies will remove the solar systems and restore the roof so it won't leak—at no cost to the customer. (Be sure you study the contractual details on this provision.)

Leases are usually fully transferable should a homeowner decide to sell. If the new buyer is interested, he or she can take advantage of the solar system on the roof.

Leases are structured to allow homeowners to make monthly payments, which helps make solar electricity a lot more affordable. In leases, customers typically pay a fixed monthly amount (kind of like budget payments made to electric companies). To determine this cost, the company simply calculates how much electricity the system should produce each year, divides it by 12, then bills you monthly for 1⁄12 of the annual total, often at a rate that is slightly below what the utility is charging.

What if you need more electricity than your PV system generates? Lease systems are grid-tied. If you need more electricity, you purchase it from your local utility. Check this carefully, as numerous utilities are now charging their solar customers a premium for electricity they purchase from them. At the end of the month, then, expect two bills—one from the solar leasing company and one from the utility.

Leases have allowed tens of thousands of homeowners to generate their own electricity from solar energy. Some companies work with schools, universities, government agencies, and corporate clients. According to their website, SolarCity has installed PV systems on "more than 400 schools, including Stanford University, government agencies such as the US Armed Forces and Department of Homeland Security, and well-known corporate clients, including eBay, HP, Intel, Walgreens and Walmart."

Representatives from the industry admit that the financial costs (to the customer) are not that different from hiring a solar company to install a system over the long haul. Bottom line, though, solar leases make it possible for any homeowner or business owner to avoid the initial cost and thus stop just talking about solar and start generating electricity from the Sun.

Community Solar

If you lack the financial ability to buy a solar system outright, or if you rent a home or apartment and can't install a system on your landlord's roof, or if you live in a home that's heavily shaded, there's yet another option: *community solar.*

Community solar started as a private venture in California, then spread to Colorado (not surprising, as these are two of the most solar-friendly states in the nation). In these early instances, the companies installed the large arrays on solar farms or the roofs of warehouses. The companies then sold modules (from one to several dozen) to subscribers in the city or town in which they were located, based on individual customer demands.

In community solar, the monthly electrical production of an individual's modules is credited to his or her utility bill. This concept, originally called *remote net metering* has spread to nearly all US states. You can find a list of solar energy leasing companies at energysage.com.

Solarize Campaigns

Community solar is a great idea, as are solar leases. However, there are other innovative ways to help make solar more affordable. If you would like a system on your home, you could possibly tap into group buying power. That is, you may be able to capitalize on volume discounts acquired by nonprofit organizations, local solar installers, and even some local governments.

One of the first projects of this nature took place in—where else?—Portland, Oregon, in 2008. Working with a nonprofit neighborhood group and the Energy Trust of Oregon, citizens banded together to purchase solar equipment in bulk. They received a volume discount that reduced the cost of their systems by 30%.

Similar projects have been undertaken in nearly 20 states in the United States. They're especially popular in New York, Connecticut, and Rhode

Island. For more information, check out the US Department of Energy's *The Solarize Guidebook,* at energy.gov.

Putting It All Together

Efficiency measures lower the size and cost of a system, often saving huge sums of money. Tax incentives and rebates also lower upfront costs. Some states exempt PV systems from sales taxes or property taxes, creating additional savings. Avoiding line extension fees by installing an off-grid system in a new home rather than a grid-connected system can also save huge amounts of money, often enough to pay for a good portion, or perhaps even all of your system cost. If buying a system isn't possible, you have several options: leases, community solar, and solar co-ops.

Economics is where the rubber meets the road. Comparing solar electric systems against the "competition," calculating the return on investment, or comparing strategies using present value gives a potential buyer a much more realistic view of the feasibility of solar energy. Just don't forget to think about *all* the opportunities to save money. If you invest in efficiency measures to lower the system cost, remember that those efficiency measures will provide a lifetime of savings, helping to underwrite your PV system. As I pointed out in Chapter 1 and again in this chapter, economics is not the only metric on which we base our decisions. Energy independence, environmental values, reliability, the cool factor, bragging rights, the fun value, and other factors all play prominently in our decisions to invest in renewable energy.

People often invest in renewable energy because they want to do the right thing. If you want to invest $12,000 to $40,000 in a PV system to lower your carbon footprint, create a better world for your children or grandchildren, or simply to live by your values, do it. It's *your* life. It's *your* money. Just know that it is also pretty profitable!

CHAPTER 5

Solar Electric Systems: What Are Your Options?

One decision you will need to make early on is the type of system you'd like. As noted previously, your options are: (1) batteryless utility-tied, aka grid-tied; (2) utility-tied with battery backup; and (3) off-grid.

In this chapter, I'll discuss each one, describe their components, how they operate, and the pros and cons of each system. I'll also discuss hybrid PV systems, that is, PV systems that work in conjunction with other renewable energy systems, such as wind.

Batteryless Utility-Tied PV Systems (aka Grid-Connected PV Systems)

The vast majority of solar electric systems installed throughout the world in more developed countries are *batteryless utility-tied systems*—so named because they connect directly to the network of electric wires that are supplied by local electric utilities (Figure 5.1). Utility networks, in turn, are connected to much larger high-voltage lines that run from state to state or province to province, often across entire nations. These transmission lines feed electricity into local electrical distribution systems and constitute the *electrical grid* (Figure 5.2a).

Even though PV systems on homes and other buildings are connected to the local utility's distribution system, as shown in Figure 5.2b, "utility-connected" or "utility-tied" PV systems are usually referred to as "grid-connected" or "grid-tied" PV systems. In this book, I'll use the term *grid-connected.*

As shown in Figure 5.3, a grid-connected system consists of five main components: (1) a PV array; (2) an inverter designed specifically for grid

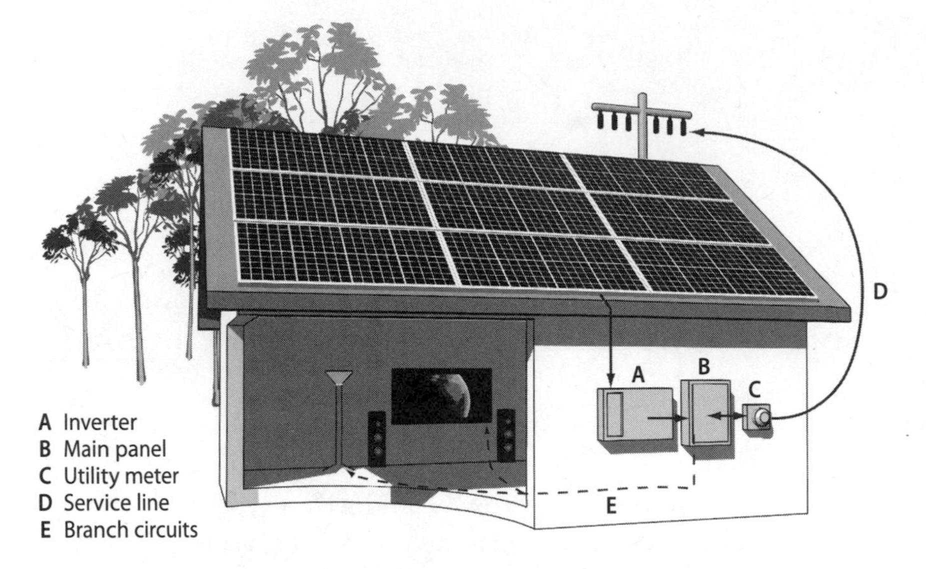

Figure 5.1. *Diagram of a simplified batteryless grid-tied PV array. Systems such as this usually consist of two or more series strings of modules, typically wired at 350 to 450 V DC. Grid-tied inverters come with an AC/DC disconnect that typically meets all the requirements of the National Electric Code. Some installers wire series strings into a combiner box (not shown), which then feeds to a DC disconnect located between the array and the inverter. Utilities commonly require an AC disconnect (not shown) between the main service panel and the utility meter.* Credit: Anil Rao.

Figure 5.2. *High-Voltage Electrical Wires.* (a) *The national electric grid consists of high-voltage wires and extremely tall towers that transmit electricity across states, allowing utilities to share electricity. Credit: Dan Chiras.* (b) *The national electric grid feeds into local utility networks that serve our homes and businesses. (Shown on p. 87).* Credit: Forrest Chiras.

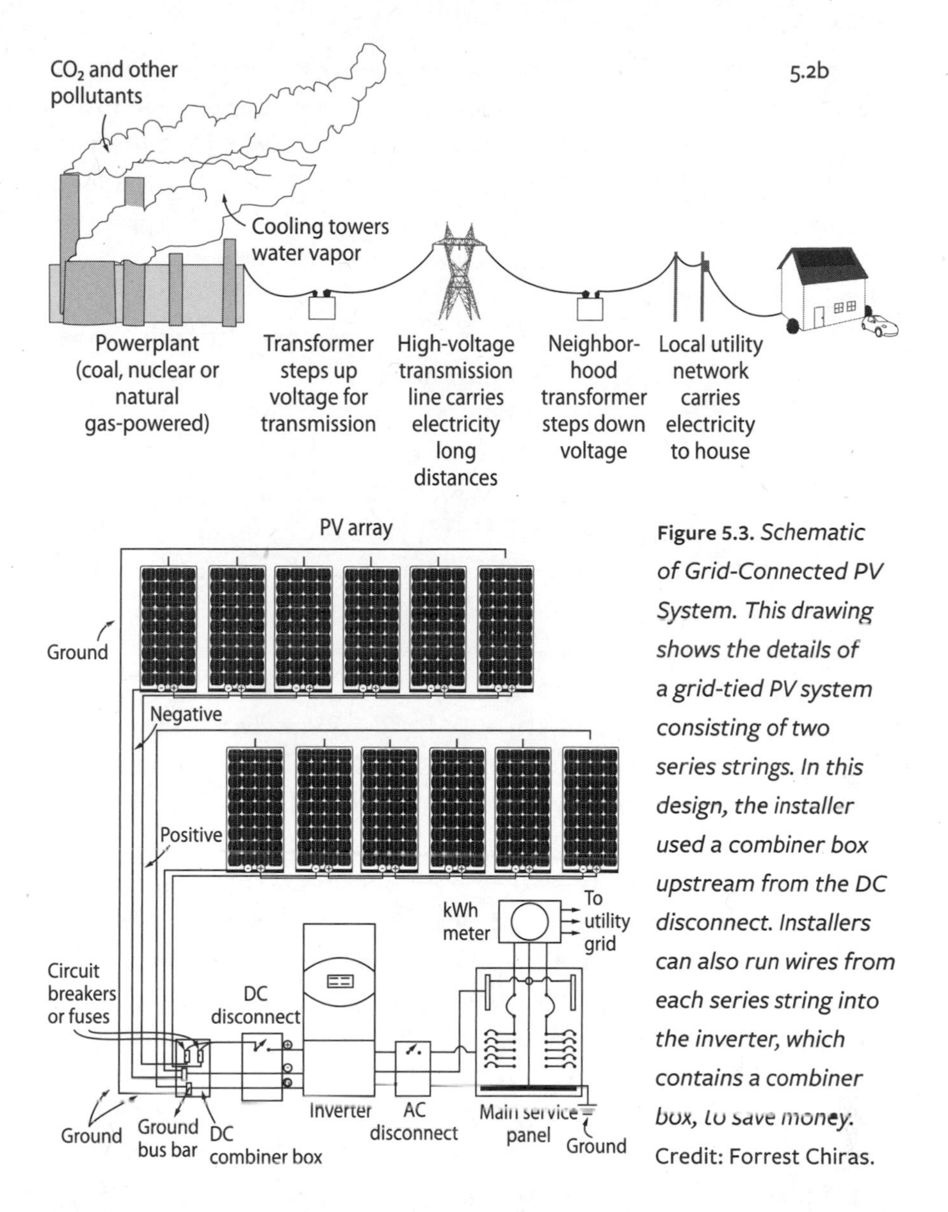

Figure 5.3. *Schematic of Grid-Connected PV System. This drawing shows the details of a grid-tied PV system consisting of two series strings. In this design, the installer used a combiner box upstream from the DC disconnect. Installers can also run wires from each series string into the inverter, which contains a combiner box, to save money.* Credit: Forrest Chiras.

connection; (3) a main service panel (or breaker box in very old homes not wired to modern Code requirements); (4) safety disconnects; and (5) a utility meter. Safety disconnects include the AC and DC disconnect. The system may also contain a DC combiner box.

To understand how a batteryless grid-connected system works, let's begin with the PV array. As you are well aware by now, PV arrays produce DC electricity; it flows through wires to an *inverter*. The inverter converts the DC electricity to AC electricity. The inverter is typically wired into one to four series strings of modules. For this reason, it is referred to as a *string inverter.* (For a description of AC and DC electricity, see the sidebar "AC vs. DC Electricity.") String inverters may be mounted indoors (usually they are installed in basements next to the main service panel) or outdoors (for example, on an exterior wall of a home or on the rack of a ground-mounted PV array).

Another inverter that's becoming increasingly popular is the *microinverter.* As shown in Figure 5.4, microinverters are miniature inverters. They are wired directly into the modules—one microinverter per module.

Microinverters convert the DC electricity produced by modules into 240-volt AC electricity. Multiple microinverters are then wired together in ways that boost the amperage. (We typically wire up to 16 microinverters in an AC series string.) Microinverters send electricity to an AC (utility) disconnect. From the AC disconnect, electricity flows directly into the main panel (Figure 5.5).

PV systems are wired into a main panel via a 240-volt circuit breaker. The National Electrical Code requires that the PV system breaker be labeled so any electrician working on the panel will know where/what it is. The main panel must also be labeled to indicate that it is powered by a second source of energy, a PV system.

Microinverter-based PV systems are popular among installers for a number of reasons. First of all, they are the simplest of all grid-tied systems to wire. In addition, PV systems containing microinverters are slightly more efficient than PV systems with string inverters. That's due in large part to the fact that they reduce the adverse effects of shading.

Figure 5.4. *Microinverter on an Array. Microinverters like the one shown here are mounted either on the module frames or the rack system in close proximity to PV modules. They convert the DC electricity into 240 V AC electricity.*
Credit: Dan Chiras.

While shading part of an array that's wired to an inverter often reduces the performance of an entire array, shading on one or two modules equipped with microinverters only affects the output of those modules. (In my opinion, this advantage isn't as huge as manufacturers suggest because all

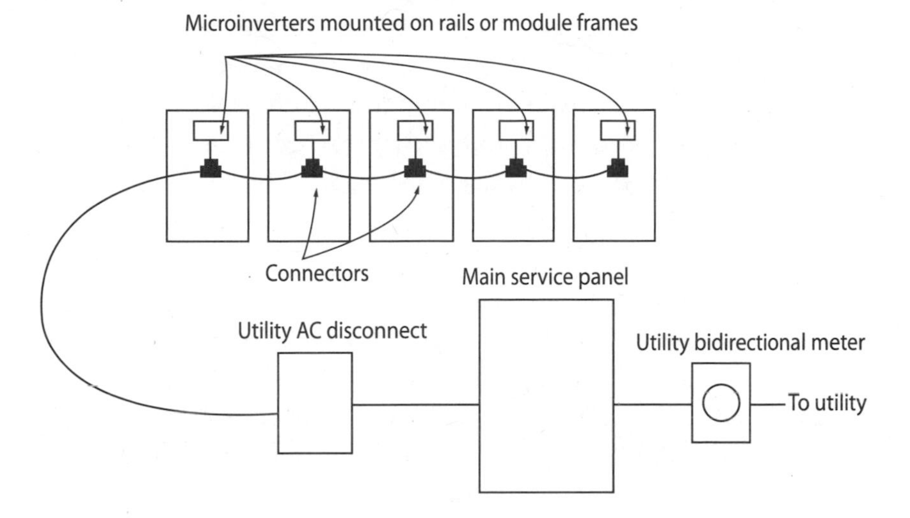

Figure 5.5. *Schematic of Grid-Connected PV System with Microinverters. This drawing illustrates the components of a grid-tied solar electric system with microinverters. In this type of array, the microinverters are wired in series, which increases the amperage of the series string. This system requires an AC disconnect, and is wired directly into the main panel.* Credit: Forrest Chiras.

AC vs. DC Electricity

Electricity comes in two basic forms: *direct current* (DC) and *alternating current* (AC). Direct current electricity consists of electrons that flow in one direction through an electrical circuit. It's the kind of electricity produced by a flashlight battery or the kind that runs electrical systems in automobiles.

Like DC electricity, AC electricity consists of the flow of electrons through a conductor. However, in alternating current, the electrons flow back and forth. That is, they change (alternate) direction in very rapid succession, hence the name "alternating current." Each change in the direction of flow (from left to right and back again) is called a *cycle.*

PV modules are designed to divert electricity around shaded portions of modules.) Another huge advantage of microinverters is that they come with a 25-year warranty.

Inverters do not just convert the DC electricity to AC, they convert it to grid-compatible AC—that is, 60-cycles-per-second, 240-volt electricity —the kind we have in our homes. (See sidebar "Frequency and Voltage" for more.) Because inverters in grid-tied systems produce electricity in sync with the grid, they are often referred to as *synchronous inverters.*

The 240-volt AC electricity produced by the inverter flows to the main service panel, aka the breaker box, the "main panel", or simply the "panel." Main panels are usually installed indoors—in basements or garages. In newer homes, main panels are typically rated at 200 amps. If you have an all-electric home, though, you may have two 200-amp panels.

From here, electricity flows through the wires in a building to active loads—that is, to electrical devices that are operating at the time. Each circuit is referred to as a *branch circuit.* If the PV system is producing more electricity than is needed to meet these demands—which is often the case on sunny days—the excess automatically flows onto the grid. We say that it is "backfed" onto the local utility network; from there, it is used by your neighbors.

Frequency and Voltage

Alternating current is characterized by two important parameters: frequency and voltage. Frequency refers to the number of times electrons change direction every second and is measured as cycles per second. (One cycle occurs when the electrons switch from flowing from point A to point B, then back to point A.) In North America, the frequency of electricity on the power grid is 60 cycles per second (also known as 60 Hertz).

The flow of electrons through an electrical wire is created by a force. Scientists refer to this mysterious electromotive force as *voltage*. The unit of measurement for voltage is *volts.* Think of voltage as electrical pressure that causes electrons to move through a conductor such as a wire. Voltage is produced by batteries in flashlights, solar electric modules, wind generators, and conventional power plants.

Electricity backfed onto the utility is "measured" by a meter mounted on the side of PV-powered homes and businesses, or sometimes on nearby electric poles (Figure 5.6.) The utility meter measures the amount of electricity delivered *to* the grid, and the amount delivered to a building *from* the grid when solar arrays are not producing electricity.

Some utilities install two meters, like the ones shown in Figure 5.6. One measures electricity backfed onto the grid by the PV system; the other measures electricity provided by the utility. Today, however, most utilities use a single digital meter, referred to as a *bidirectional meter* (Figure 5.7).

Figure 5.6. *Utility Meter Mounted on Pole. In most homes, utility meters are mounted on the side of the building near the main service entrance—where the electricity enters the home from the utility. Utility meters may also be mounted on poles near a home or business. In this system, two meters were installed, one to track electricity from the grid and the second to track solar electricity fed back onto the grid. These meters are read remotely by the utility over the electrical line.* Credit: Dan Chiras.

Figure 5.7. (a) *Dial-type and* (b) *Digital Electric Meters. These photos illustrate the two types of utility meter commonly encountered in solar installations. In older homes, dial-type analog electric meters are frequently found; utilities now usually switch them out for digital electric meters.* Credit: Dan Chiras.

Grid-connected solar electric systems also contain two safety disconnects. They are shown schematically in Figures 5.3 and 5.8. Safety disconnects are manually operated switches that enable electricians to terminate the flow of electricity at key points in a PV system to prevent electrical shock when servicing a PV system. The National Electrical Code (NEC) requires a disconnect in the DC wiring (wires from the array to the inverter) and the AC wiring (wires running from the inverter to the main panel). All grid-tied inverters have a built-in AC/DC disconnect that meets this requirement.

Although the AC/DC disconnect on the string inverter meets Code requirements, most building departments require grid-connected systems to also contain a manual AC disconnect. This is known as a *utility disconnect.* It allows utility workers to disconnect a PV system from the electrical grid should they need to work on an electric line. Its purpose is to protect utility workers from shock caused by electricity backfed onto the grid by a PV system.

Shown in Figure 5.9, the utility AC disconnect must be mounted outside, typically in close proximity to the building's utility service entrance (where electrical wires enter a building).

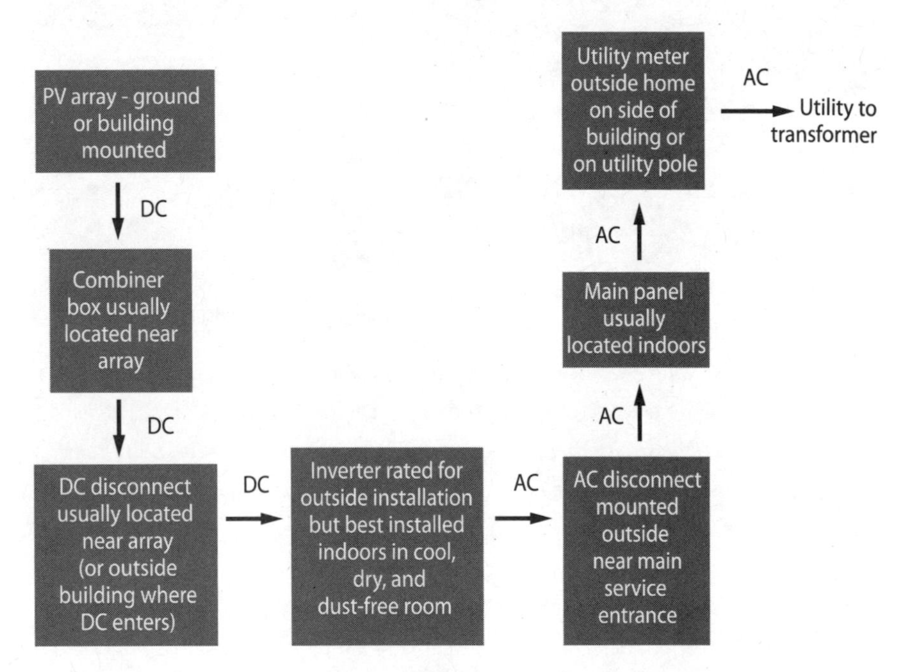

Figure 5.8. *AC and DC Disconnect Placement. This diagram illustrates the proper placement of an AC and DC disconnect in systems.* Credit: Forrest Chiras.

Figure 5.9. *Label on Utility Disconnect. Utility disconnects like the one shown here must, according to the National Electric Code, be properly labeled to warn electricians that even though the switch is open, the terminals can be live (right panel). Notice the surge protector on the side of the disconnect (left panel) and the hole in the switch handle through which the utility inserts its lock should they need to lock a system out of the utility network for repair. Utilities typically require installers to mount the AC disconnect close to the utility meter.* Credit: Dan Chiras.

The NEC requires that all AC disconnects be clearly labeled "Utility Disconnect" or "Interconnection Disconnect Switch" using a label that will withstand all weather conditions and UV light (Figure 5.9).

Although utility disconnects are required by most utilities, large utilities in California and Colorado (that have thousands of solar electric systems online) have dropped this requirement. That's because string inverters and microinverters will automatically shut down if the grid goes down. Grid-tied PV systems will not backfeed onto a dead grid. Period.

Inverters primarily shut down under two conditions: brownouts and blackouts. A brownout occurs when electrical demand is extremely high, for instance, on a hot summer day when everyone's running an air conditioner. In such instances, voltage and frequency in the line may drop. The inverter senses this perturbation, and shuts down. During blackouts there is a complete loss of electricity due to a downed line or a transformer that's been struck by lightning. The inverter senses this and shuts down automatically.

Rapid Shut Down

When a fire occurs in a home or business, the first firefighters to arrive on the scene pull the utility meter. This terminates the flow of electricity from the grid to a building. All grid-tied solar systems shut off immediately. However, if a fire occurs during the daytime, the DC wires running from the array to the inverter remain live. There's no current in them, but they have voltage. If severed by firefighters, for example, cutting through a roof to douse an attic fire, the DC wires (between the array and the inverter) could produce potentially lethal shocks.

To avoid this problem, the National Electrical Code requires the installation of a *rapid shutdown switch* or *RSS* (Figure 5.10a). It enables

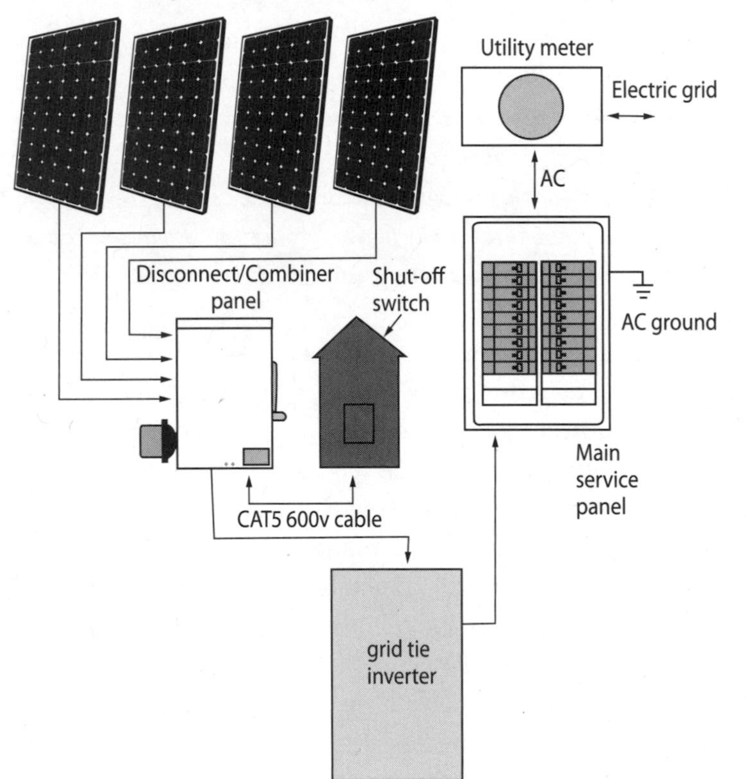

Figure 5.10a. *Rapid Shutdown Switch. One of the newest requirements of the National Electric Code is the rapid shutdown switch. It is a safety measure lobbied for by the firefighters of the United States to provide additional protection in case of a house fire.* Credit: Forrest Chiras.

Figure 5.10b. *Rapid Shutdown Switch and DC Disconnect. The 2014 National Electric Code requires a rapid shutdown that is manually or automatically activated.* Credit: MidNite Solar.

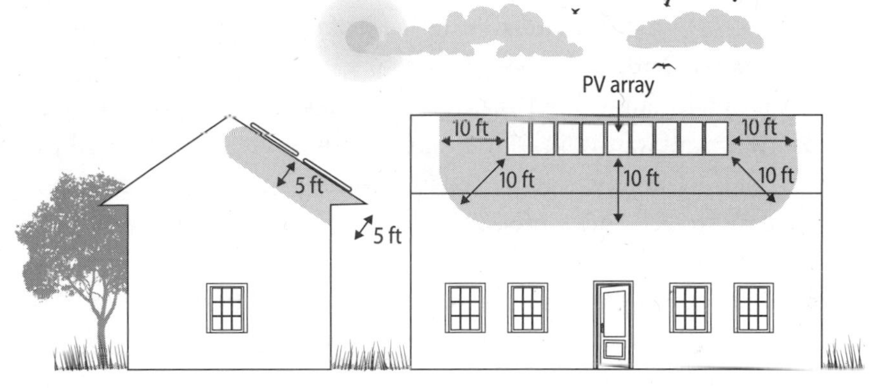

Figure 5.11. *This drawing shows the unprotected zones.* Credit: Forrest Chiras.

firefighters to terminate the flow of DC electricity within 10 feet (3.3 meters) of the array in installations where DC wires run across the surface of a roof or within 5 feet (1.6 meters) of DC wires entering a building, as illustrated in Figure 5.11.

As illustrated in Figure 5.10a, the rapid shutdown mechanism consists of two components: a shutoff switch located near the service entrance and a DC disconnect panel (labeled disconnect/combiner panel). When firefighters arrive, they pull the utility meter and then switch off the RSS. (Some RS switches turn off automatically.) Figure 5.10b shows what the equipment looks like. Systems containing these components are designed to terminate the flow of electricity in wires running on the surface of the roof within 10 feet of an array and within 5 feet of the array if the wires run inside the building, as illustrated in Figure 5.11.

RSS requirements can add a couple thousand dollars to a solar electric system installation (for equipment and labor), but they are only required in systems with string inverters. Microinverter systems don't require them.

Net Metering and Billing in Grid-Connected Systems

All customers who connect their PV systems to the grid enter into a contractual agreement with their utility. Called an *interconnection agreement,* it spells out many details, including the provisions for paying a customer for surplus. Surplus electricity is referred to as *net excess generation.* This payment language is part of a net metering policy established by the state. Nearly every state has one.

Two types of net metering policies exist for handling net excess generation (NEG): monthly and annual. Let's start with annual, as it is a bit easier to understand.

In states with annual net metering, utilities carry surplus electricity (measured in kilowatt-hours) from one month to the next for a full year. Because of this, surpluses generated in summer months can make up for shortages in the fall and winter. What happens to surpluses at the end of the year?

In annual net metering, unused surplus electricity remaining in the account at the end of the year can be handled in one of three ways. It is: (1) transferred to the utility (forfeited); (2) purchased by the utility at the retail price of electricity, i.e., the same price that a customer pays the

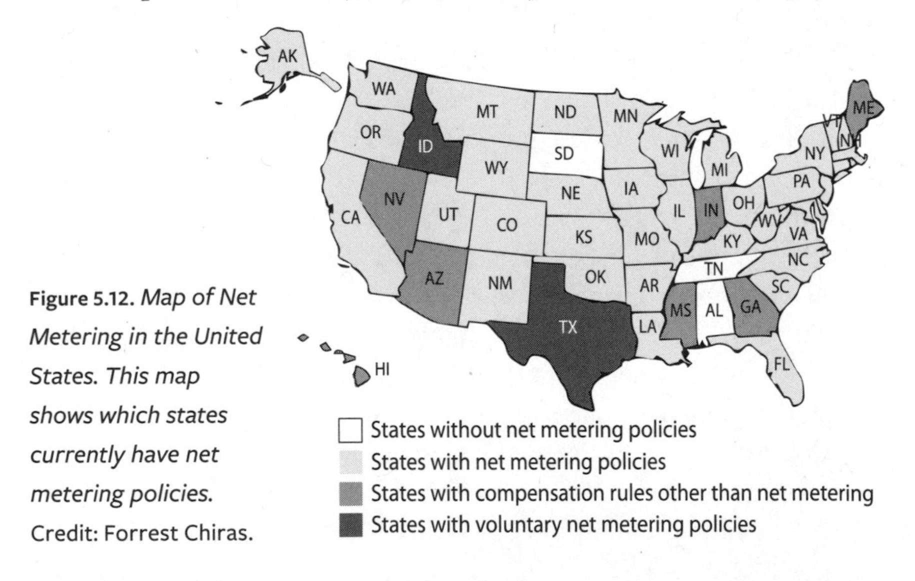

Figure 5.12. *Map of Net Metering in the United States. This map shows which states currently have net metering policies.* Credit: Forrest Chiras.

utility for electricity; or (3) purchased by the utility at its wholesale rate. Wholesale rate is referred to as *avoided cost.* It's what electricity costs that utility—either to make it or buy it from another supplier. It's usually about one-fourth the cost of electricity (retail price) to customers.

Once the surplus is reconciled, the balance is set to zero, and the net metering starts over for another year. That said, some states (such as extremely renewable-energy-friendly Colorado) allow customers to carry their balance from year to year. Thus, if you have a surplus in 2020, you can carry it over to 2021 or 2022. If you need it in 2021 or 2022, it's yours for free.

The advantage of annual net metering is that it accommodates the seasonal variation in a PV system's production. In the summer months in many climates, PV systems produce more electricity than is consumed by a customer. In the winter months, PV systems typically produces less electricity than is consumed. Shortfalls during the winter can be met by drawing summer surpluses from the "bank."

In states with monthly net metering, utilities deal with the surpluses, if any, at the end of each month. This arrangement is known as *monthly net metering.* In monthly net metering, surpluses are carried from day to day in each month. If your home generates a surplus on Monday, you can have it back for free on Tuesday or even the following Monday, as long as that date falls within the monthly billing period. However, any surpluses remaining at the end of the month must be reconciled. They can't be carried over. Utilities typically credit surpluses to the customers' next bill and credit it at avoided cost or the full retail value.

If you're thinking that connecting to the grid could be profitable venture, don't get your hopes up. There aren't many states that reimburse customers at retail rates for monthly net excess generation. Most utilities pay for surpluses at the avoided cost the cost of generating power. In Missouri, for instance, its monthly net metering policy allows utilities to reimburse at avoided cost. Although you may be paying the utility 10 or 12 cents per kilowatt-hour, they'll only pay you about 2.5 cents. Some states like Arkansas simply "take" the surplus without payment to the customer—it all depends on state law. (If you're not happy with your state law, consider working to change it!)

Monthly net metering is generally the least desirable option, especially if surpluses are "donated" to the utility company or reimbursed at avoided cost. Annual reconciliation is a much better deal. The ideal arrangement, from a customer's standpoint, is a continual rollover. In such instances, there's no concern about losing your banked solar-powered kilowatt-hours. (I've found 13 states that permit continued rollover.)

Net metering is mandatory in many states, thanks to the hard work of renewable energy activists and forward-thinking legislators. At this writing (February 2019), net metering policies exist for utilities in 47 states and the District of Columbia. Only three states have no net metering policy at this writing: Alabama, Tennessee, and South Dakota. Even so, three utilities in these states have adopted net metering. Texas and Idaho have voluntary programs, which means the utilities can do whatever they want. Table 5.1 lists the states with the best policies.

Although virtually all states have passed net metering policies, there are major differences. Differences include (1) who is eligible, (2) which utilities are required to participate, (3) the size and types of systems that qualify, and (4) reimbursement for NEG. For more on net metering, you may want to pick up a copy of my book, *Power from the Sun.*

Although they are reluctant to admit it, many utilities are finding that PV systems actually help them meet demand and do so quite economically. That's because residential and business PV systems often help "shave" peak load—that is, reduce daily electrical demand during periods of high use. As you can see in Figure 5.13, peak demand runs from about 11 a.m. to 6 p.m. To meet electrical demand during this time, many utilities buy power from other utilities or fire up additional generating capacity, such as natural-gas-powered electric plants. Both options can be very costly. Purchasing on the "spot market" can be extremely expensive. While a utility may generate electricity for 3 cents per kilowatt-hour, electricity purchased on the spot market (from other sources) can cost as much as 25 cents per kWh. Solar electricity is often generated in surplus while families are away at work or school, so PV systems can help utilities reduce costs, saving the

Table 5.1. States with the Most Favorable Net Metering Policies

Arizona	Delaware	New Jersey	Pennsylvania
California	Maryland	New York	Utah
Colorado	Massachusetts	Ohio	Vermont
Connecticut	New Hampshire	Oregon	West Virginia

Source: "Best and Worst Practices in State Net Metering Policies and Interconnection Procedures," Freeingthegrid .org.

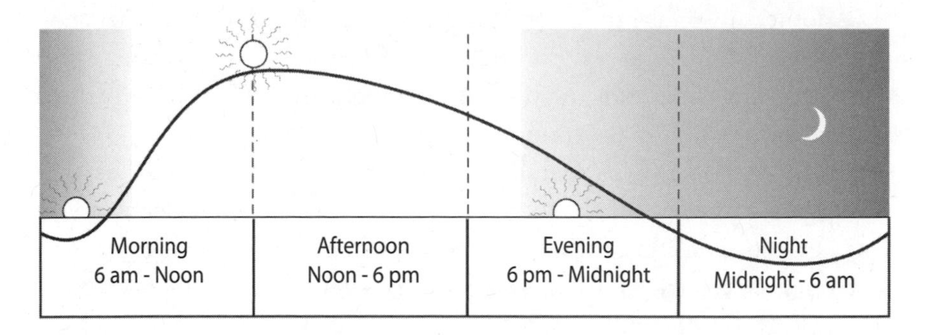

Figure 5.13. *Peak Demand. Peak load, that is, peak electrical demand, begins late in the morning and continues until the early evening in most regions of the United States. Solar electric systems help utilities shave peak demand, that is, reduce the demand for electricity during this time. This, in turn, makes it easier for utilities to supply inexpensive electricity to their customers.* Credit: Anil Rao.

utility and customers money. Even if a utility reimburses PV customers for net excess generation at retail—say 12 cents/kWh—they are avoiding purchase on the spot market at 25 cents. It's a good deal for them.

The Pros and Cons of Grid-Connected Systems

Grid-connected PV systems have their pluses and minuses, summarized in Table 5.2. On the positive side, batteryless grid-connected systems are relatively easy to install. They are simpler than other PV systems and are, therefore, considerably less expensive—often 30%–40% cheaper than off-grid systems and about 20%–30% cheaper than grid-connected systems with battery backup.

Grid-tied systems also avoid costly battery replacement—something you can count on every seven to ten years if you have a battery-based system. No batteries means very little, if any maintenance.

Table 5.2. Pros and Cons of Batteryless Grid-Tie Systems

Pros	Cons
Simpler than other systems	Vulnerable to grid failure unless a backup generator or an uninterruptible power supply is installed
Less expensive	
Less maintenance	
Unlimited supply of electricity (unless the grid is down)	
More efficient than battery-based systems	
Unlimited storage of surplus electricity (unless the grid is down)	
Greener than battery-based systems	

Another advantage of batteryless grid-connected systems is that they can store months of excess production—if you are tied to a utility that offers annual net metering. As you can imagine, they don't physically store surplus, they "store" in the form of a credit on your utility bill.

The grid also serves as an unlimited battery bank. Unlike a real battery bank, you can never "fill up" the grid. It will accept as much electricity as you can feed it. In contrast, when batteries are full, they're full. They can't take on more electricity. As a result, surpluses generated in an off-grid PV system are typically lost.

Another advantage of grid-connected systems is that the grid is always there—well, almost always. Even if the Sun doesn't shine for a week, you have access to a reliable supply of electricity. With an off-grid system, at some point the batteries may run down. When they do, you need to run a backup generator—or sit in the dark.

Yet another advantage of grid-connected systems with net metering is that utility customers suffer no losses when they store surplus electricity on the grid. For reasons described in Chapter 7, when electricity is stored in a battery, as much as 20% to 30% of the electrical energy fed into a battery bank is lost. If you deliver 100 kWh of electricity to the grid, however, you can draw off 100 kWh. (This assumes annual net metering with the year-end reconciliation at retail price.) I'd be remiss if I didn't point out the grid has losses too, but net-metered customers get 100% return on their stored electricity.

Another advantage of batteryless grid-tie systems is that they are greener than battery-based systems. Although utilities aren't the greenest businesses in the world, they are arguably greener than battery-based systems. That's because batteries require an enormous amount of energy to produce. Lead must be mined and refined. Batteries must be assembled and shipped. Batteries contain highly toxic sulfuric acid and lead. Although old lead-acid batteries are recycled (there's at least one company in the United States [Deka] that recycles its own batteries), they're often recycled under less-than-ideal conditions. Many companies ship batteries to less-developed countries, where children are often employed to remove the lead plates along the banks of rivers. Acid from batteries may contaminate surface waters.

When annually net-metered at retail rates, batteryless grid-connected systems provide substantial economic benefits. As noted in Chapter 1, a grid-tied solar electric system produces electricity at a cost much lower than utilities.

On the downside, batteryless grid-connected PV systems are vulnerable to grid failure. If the grid goes down, so does your PV system. Even if the Sun is shining, these systems will shut down if the grid experiences a brown-out or blackout. If power outages are a recurring problem in your area and you want to avoid service disruptions, you may want to consider installing a backup generator that automatically switches on if the grid goes down.

Or, you could install an uninterruptible power supply (UPS) on critical equipment, such as computers. A UPS consists of a battery pack and an inverter. If the utility power goes out, the UPS will supply uninterrupted power until its battery gets low.

A third option, discussed in the next section, is installing a grid-connected system with battery backup. In this case, batteries provide backup power during a power outage. There's also an inverter that will provide limited power if the grid goes down, which I'll describe in Chapter 6.

Grid-Connected Systems with Battery Backup

If you want to be connected to the utility but would like backup power in power outages, a grid-connected system with battery backup is what you'll need. These are known by several tongue-twisting names: *battery-based utility-tied* systems or *battery-based grid-connected* systems. Grid-connected systems with battery backup ensure a continuous supply of electricity, even when an ice storm wipes out the electrical supply to you and huge swaths of your utility company's service area.

These systems contain all of the components found in grid-connected systems: a PV array; an inverter; AC and DC disconnects, including a rapid shutdown switch; a main service panel; and a utility meter (or two) to keep track of electricity delivered to and drawn from the grid (Figure 5.14). However, there are a few key differences.

The first difference is the inverter. (They're designed to function with batteries and the grid.) And, of course, battery-based grid-connected systems also require battery banks. Battery banks in these systems, however, are typically small—one-third to one-fourth the size of a battery bank required for an off-grid system. Small battery banks are needed because these systems are designed to provide sufficient storage to run a few critical loads for a day or two while the utility company restores electrical service.

Critical loads might include a few lights, the refrigerator, a well pump, the blower of a furnace, the pump in a gas- or oil-fired boiler, and a sump pump. Those who want or need full power during outages must install much larger and costlier battery banks or backup generators.

In grid-tied systems with battery backup, batteries are called into duty only when the grid goes down. They're strictly there to provide backup power. They're not installed to supply additional power, for example, to run loads that exceed the PV system's output. That's the job of the grid. Because you want a full battery banks if the grid goes down, batteries are maintained at full charge day in and day out.

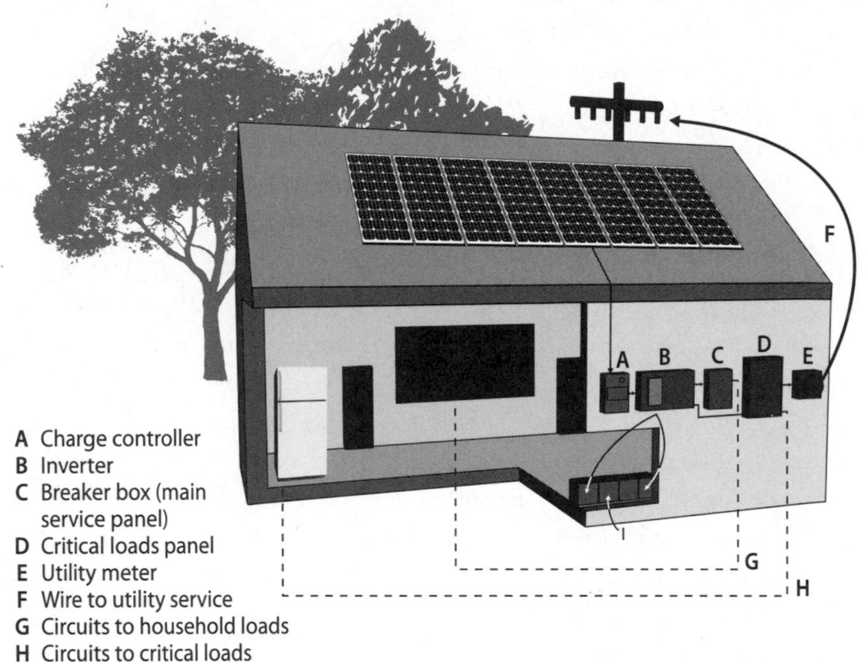

Figure 5.14. *Grid-Connected System with Battery Backup. This drawing shows the main components of a grid-tied system with battery backup. Please note that it does not show DC and AC disconnects or a combiner box. It does show the main service panel and the critical loads panel, however. In these systems, installers have historically wired modules in very short series strings because of voltage limitations of charge controllers. Newer, high-voltage charge controllers allow installers to wire series strings in much higher voltages, increasing efficiency, reducing wire size, and potentially saving homeowners a little money.* Credit: Forrest Chiras.

Maintaining a fully charged battery bank requires a tiny amount of energy over long periods. That is to say, a small portion of the electricity a PV system generates is devoted to keeping batteries full at all times. This reduces overall system efficiency by 5% to 10%.

Battery banks in grid-connected systems don't require the careful monitoring like those in off-grid systems do, but it is a very good idea to keep a close eye on them. When an ice storm knocks out your power, the last thing you want to discover is that your battery bank quietly died on you last year. I'll explain battery monitoring in Chapter 7.

Another key component of grid-connected systems with battery backup is the *charge controller,* shown in Figures 5.14 and 5.15. Although I'll discuss charge controllers in depth in Chapter 7, a brief discussion is helpful at this point.

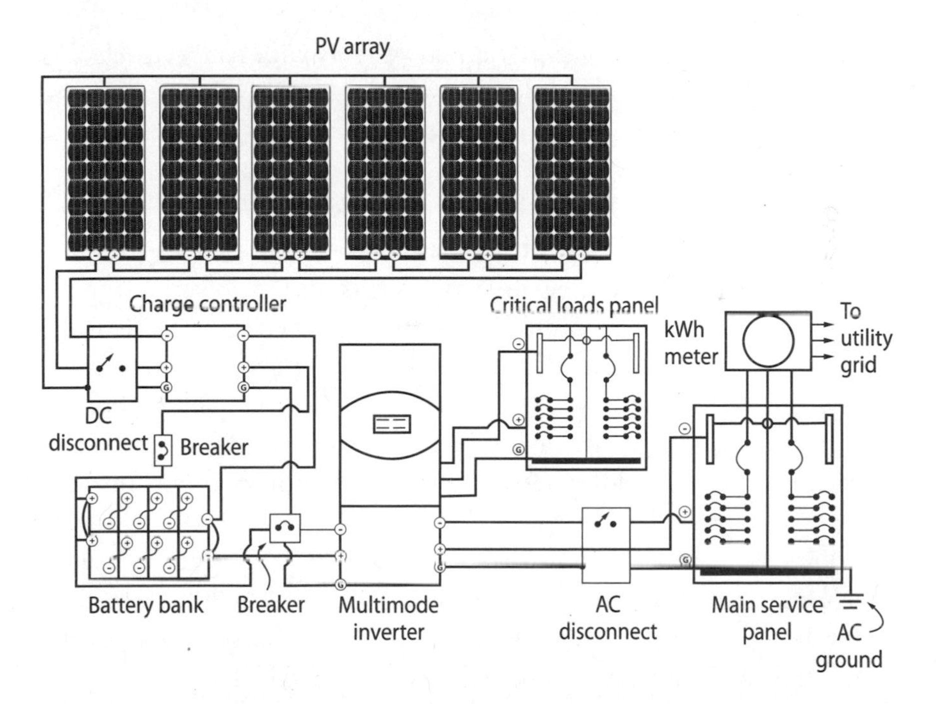

Figure 5.15. *Schematic of Grid-Connected System with Battery Backup. In these systems, the batteries are kept at full charge. Batteries are not called into duty unless there is a power outage. In that case, the system converts seamlessly to battery power. When this occurs, the inverter stops sending electricity to the main service panel, which prevents electricity from being backfed onto the grid. It only delivers electricity to the critical loads panel.* Credit: Forrest Chiras.

The charge controller performs several vital functions. For example, it regulates the flow of electricity into a battery bank when it is being charged in a way that ensures your batteries are rapidly and efficiently charged. Charge controllers also prevent batteries from overcharging. I'll explain how in Chapter 7. Just remember this: overcharging can permanently damage lead-acid batteries.

Charge controllers also protect batteries discharging too deeply, referred to as "over discharging." This, too, can seriously damage batteries. I'll discuss this in Chapter 7 as well.

Modern charge controllers for residential solar systems come with a feature called *maximum power point tracking* (MPPT). Discussed in Chapter 7, MPPT optimizes the output of a PV array, thus ensuring the highest possible output at all times. This improves the PV system's performance by about 15%.

Another key component of grid-tied systems with battery backup is the *critical loads panel,* as shown in Figures 5.14 and 5.15. It's a separate breaker box that services critical loads.

Pros and Cons of Grid-Connected Systems with Battery Backup

Grid-connected systems with battery backup enable homeowners and businesses to continue to operate critical loads during power outages. A homeowner, for instance, could run a refrigerator and some lights while neighbors grope around in the dark and the food in their refrigerators spoils. These systems allow businesses to continue to use computers and other vital electronic equipment so they can continue operations while their competitors twiddle their thumbs and complain about financial losses.

Although battery backup may seem like a desirable feature, it does have some drawbacks. For one, these systems cost 20%–30% more than batteryless grid-tied systems. The higher cost, of course, is due to the installation of additional components, including the charge controller, critical loads panel, and batteries. Flooded lead-acid batteries and sealed batteries used in these and other renewable energy systems are also rather expensive. A battery bank also needs a safe, comfortable home. If you are building a new home or office, you will need to add a well-ventilated battery box or a special vented battery room that stays warm in the winter and cool in the summer just to house your batteries. Battery banks generally need to be vented to the outside to prevent potentially dangerous hydrogen gas buildup, which can lead to explosions and fires if ignited by a spark or flame. Vented battery rooms and battery boxes add expense.

Flooded lead-acid batteries also require periodic maintenance and replacement. As explained more fully in Chapter 7, to maintain batteries for long life, you'll need to monitor fluid levels regularly and fill batteries with distilled water every few months. Because battery banks in these systems are infrequently used, they tend to be forgotten. Out of sight, out of mind. If a homeowner fails to maintain batteries for several years, the electrolyte level can drop below the top of the lead plates, causing irreversible damage.

Even if well maintained, battery banks need to be replaced periodically. Typical batteries used in this system require replacement every five to ten years, at a cost of around two to three thousand US dollars each time. (Batteries in grid-tied systems with battery banks tend to need replacement more often than batteries in off-grid systems.)

Yet another problem, noted earlier, is that battery-based grid-tie systems consume a portion of the daily renewable energy production just to keep the batteries topped off (fully charged).

When contemplating a battery-based grid-tied system, you need to ask yourself three questions: (1) How frequently does the grid fail in your area? (2) What critical loads are present and how important is it to keep them running? (3) Are there other less costly options, like a backup generator?

If the local grid is extremely reliable, you don't have medical support equipment to run, your computers aren't needed for business or financial transactions, and you don't mind using candles on the rare occasions when the grid goes down, why buy, maintain, and replace costly batteries? See Table 5.3 for a quick summary of the pros and cons of battery-based grid-connected systems.

Table 5.3. Pros and Cons of Battery-Based Grid-Tie System

Pros	Cons
Provide a reliable source of electricity	More costly than batteryless grid-connected systems
Provide emergency power during a utility outage	Less efficient than batteryless grid-connected systems
	Less environmentally friendly than batteryless systems
	Require more maintenance than batteryless grid-connected systems

Off-Grid (Stand-Alone) Systems

Off-grid systems are designed for individuals who want to or must supply all of their needs via solar energy—or a combination of solar and some other renewable source. As shown in Figure 5.16, off-grid systems bear a remarkable resemblance to a grid-connected system with battery backup. There are some noteworthy differences, however. The most important is that there *is* no grid connection. As you can see in Figure 5.16, there are no power lines running from the house or business to the grid. These systems "stand alone." The main source of electrical energy in an off-grid system is the PV array. Electricity flows from the PV array to the charge controller. The charge controller delivers DC electricity to the battery bank. When electricity is needed, it is drawn from the battery bank. The inverter converts the DC electricity from the battery bank, typically wired at 24 or 48 volts, to higher-voltage AC, typically the 120 volts required by off-grid households and businesses. AC electricity then flows to active circuits in the house via the main service panel.

Off-grid systems often require a little "assistance" in the form of a wind turbine, microhydro turbine, or a gasoline or diesel generator, often

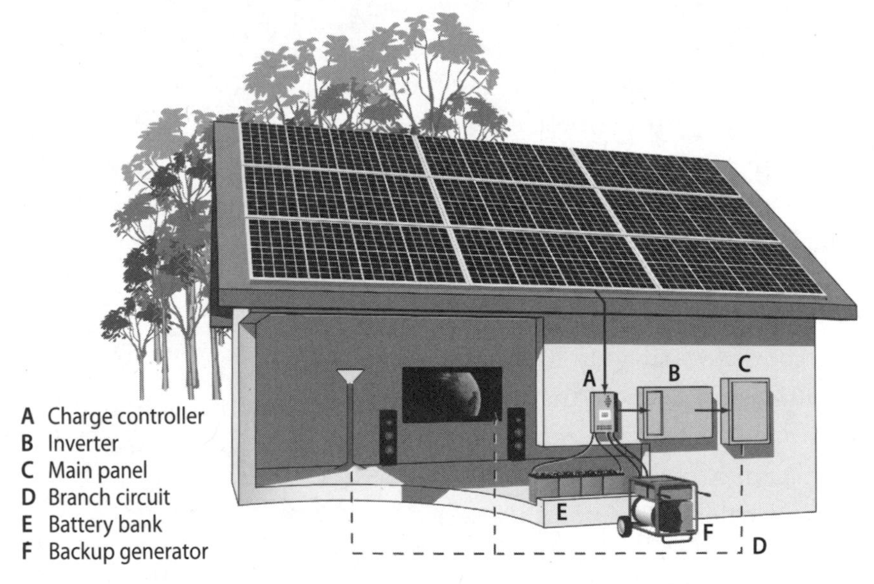

Figure 5.16. *Off-Grid System. Off-grid solar electric systems typically require a combiner box (not shown here), a charge controller, an inverter/charger, and a main service panel. A backup generator is often installed for battery maintenance and to provide power during unusually cloudy periods.* Credit: Anil Rao.

referred to as a *gen-set*. One or more of these energy sources helps make up for shortfalls. The majority of off-grid systems include generators. "A gen-set provides redundancy," notes National Renewable Energy Laboratory's wind-energy expert Jim Green. Moreover, "if a critical component of a hybrid system goes down temporarily, the gen-set can fill in while repairs are made." Finally, gen-sets also play a key role in maintaining batteries, a subject discussed in Chapter 7.

Because most backup generators produce alternating current electricity, it must be converted to DC to charge the batteries. This function is controlled by a battery charger located in the off-grid inverter. That's why inverters in battery-based systems are referred to as *inverter/chargers.*

Like grid-connected systems with battery backup, an off-grid system requires safety disconnects to permit safe servicing. A DC disconnect should be located between the PV array and the charge controller, and an

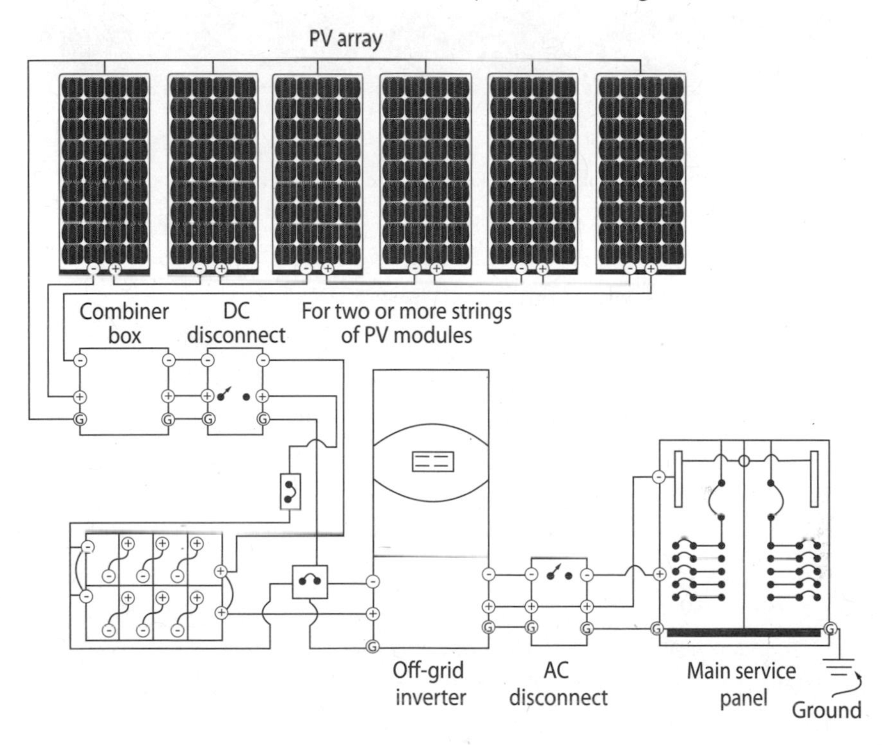

Figure 5.17. *Schematic of Off-Grid System. This drawing shows the main components of an off-grid solar electric system, including the combiner box when two or more series strings of PVs are installed, the DC disconnect, charge controller, backup generator, and battery bank.* Credit: Forrest Chiras.

AC disconnect should be installed between the inverter and the main service panel. A sign must be posted to warn firefighters that a home contains an off-grid PV system with a battery bank.

Off-grid systems also require charge controllers to protect the batteries, as explained in the previous section. Charge controllers also contain the maximum power point tracking circuitry mentioned earlier. Charge controllers not only prevent batteries from being overcharged, they also prevent reverse current flow from the battery back to the array at night. Although reverse current flow is typically very small, it is best avoided. Modern-day charge controllers make this a non-issue.

As is evident by comparing schematics of the three types of systems, grid-tied systems with battery backup are the most complex. Second in terms of complexity is the off-grid system.

Why Wire a Home for DC?

Most modern homes and businesses operate on alternating current electricity. However, off-grid homes supplied by wind or solar electricity—or a combination of the two—can be wired to operate partially or entirely on direct current electricity to power DC lights, refrigerators, televisions, and even ceiling fans. Why wire a home or cottage for DC?

One reason is that DC systems do not require inverters. Electricity flows directly out of the battery bank to service loads. This, in turn, can reduce the cost of the system by $2,000 to $4,000. That said, it is important to note that cost-savings created by avoiding an inverter may be reduced by higher costs elsewhere in the system. For example, DC appliances usually cost more than AC appliances—considerably more. DC ceiling fans, for instance, cost two to six times more than comparable AC models. You could pay $300–$350 for a DC model, but $50–$150 for a comparable AC ceiling fan.

DC appliances and electronics are not only more expensive, they are more difficult to find. You won't find them at national or local appliance and electronics retailers. Many DC appliances are tiny, too. Most DC refrigerators, for example, are miniscule compared to the AC models used in homes. That's because DC appliances are primarily marketed to boat and recreational vehicle enthusiasts, and there's not a lot of room in a boat or recreational vehicle for large appliances. In addition, DC refrigerators and freezers do not come with the features that many people expect, such as automatic defrost, ice makers, or cold-water dispensers on the door. They also cost much more than AC units, even high-efficiency models. Because of these reasons, I rarely recommend the inclusion of DC circuits for off-grid homes. They are, however, often installed in remote cabins and cottages that are only used occasionally.

Pros and Cons of Off-Grid Systems

Off-grid systems offer many benefits, including total emancipation from the electric utility (Table 5.4). Off-grid systems also provide freedom from power outages.

Off-grid systems do have some downsides. First, they are the most expensive of the three systems. Large battery banks and backup generators add substantially to the cost—often 50% to 60% more. In addition, they require space to house battery banks and generators. Batteries also require periodic maintenance and replacement every seven to ten years, depending on the quality of batteries you buy and how well you maintain them.

Although cost is usually a major downside, there are times when off-grid systems cost the same or less than grid-connected systems—for example, if a home or business is located more than a few tenths of a mile from the electric grid. Under such circumstances, it can cost more to run electric lines to a home than to install a small off-grid system.

When installing an off-grid system, remember that you become the local power company, and your independence comes at a cost to you. You will very likely need to buy a gen-set. Gen-sets cost money to maintain and operate, and you may be dependent on your own ability to repair your power system when something fails.

Independence also comes at a cost to the environment.

Table 5.4. Pros and Cons of an Off-Grid System

Pros	Cons
Provide a reliable source of electricity	Generally the most costly solar electric systems
Provide freedom from utility grid	Less efficient than batteryless grid-connected systems
Can be cheaper to install than grid-connected systems if located more than 0.2 miles from grid	Require more maintenance than batteryless grid-connected systems (you take on all of the utility operation and maintenance jobs and costs)

Gen-sets produce air and noise pollution. Lead-acid batteries are far from environmentally benign, as noted earlier in this chapter. So, think carefully before you decide to install an off-grid system.

Off-grid systems also require a huge commitment to energy efficiency and a major change in lifestyle. You won't be able to install devices that consume a lot of electricity such as electric furnaces, electric water heaters, conventional hot tubs, or geothermal systems to heat and cool your home. They use way too much energy for an off-grid system. You'll need to eliminate all phantom loads as well and be especially vigilant about leaving lights and electronic equipment running when not in use. Every appliance or electronic device you purchase will need to be as energy efficient as possible. In short, the average energy-wasteful lifestyle is not amenable to an off-grid system.

Hybrid Systems

As you've just seen, you have three basic options when it comes to solar electric systems. Each of these systems can be designed to include additional renewable energy sources. The result is known as a *hybrid renewable energy system* (Figure 5.18).

Hybrid renewable energy systems are extremely popular among homeowners in rural areas. One reason is that solar electricity and wind are a marriage made in heaven in many parts of the world. Why?

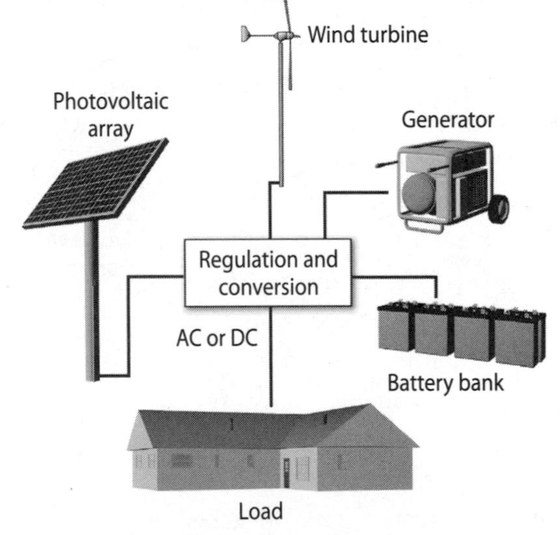

Figure 5.18. *Hybrid PV and Wind System. Solar electric and wind energy systems can be installed along with a backup generator, if necessary. This is a simplified view of the system. Wind turbines and solar electric systems are generally wired to their own inverters, although there are a few inverters that can be wired directly to a hybrid solar electric and wind system.* Credit: Anil Rao.

In most locations, solar energy and winds vary throughout the year. Solar radiation striking the Earth tends to be highest in the spring, summer, and early fall. Winds tend to be strongest in the late fall, winter, and early spring. This complementary relationship is shown graphically in Figure 5.19.

In areas with sufficient solar and wind resources, a properly sized hybrid PV/wind system can not only provide 100% of your electricity, it may eliminate the need for a backup generator. A wind system may also be used to maintain batteries, which will help you extend the lifespan of your batteries, discussed in Chapter 7.

Because wind and PVs complement each another, you may be able to install a smaller solar electric array and a smaller wind generator than if either were the sole source of electricity. Bear in mind, to make wind work for you you'll need an excellent location and a very tall tower for your turbine—typically 100 to 120 feet (30 to 36 meters) high!

If the combined solar and wind resource is not sufficient throughout the year or the system is undersized, a hybrid system will require a backup generator to supply electricity during periods of low wind and low sunshine. Gen-sets are also used to maintain batteries in peak condition and permit use of a smaller battery bank.

Despite the benefits of hybrid systems, my experience strongly suggests that rather than installing two separate systems, you'd be a lot better

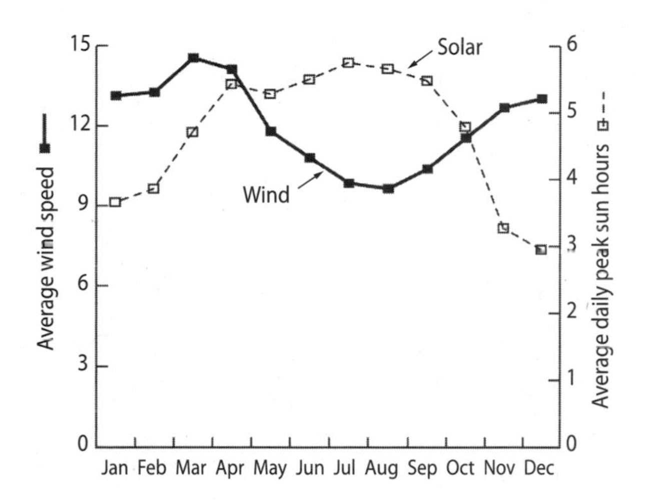

Figure 5.19. *Complementary Nature of Wind and Solar. This graph of solar and wind resources at my educational center, The Evergreen Institute, and home in Gerald, Missouri, show how marvelously complementary wind and solar energy are. For off-grid systems, a combination of solar and wind can ensure a reliable amount of electricity year round. Be careful, however, as it may make more sense to simply expand a solar electric system.* Credit: Anil Rao.

off simply expanding your solar electric system. It's far cheaper to install a larger solar electric system than to install a wind turbine to complement a solar electric system.

Choosing a PV System

To sum things up, you have three basic choices when it comes to a PV system. If you have access to the electric grid, you can install a batteryless grid-connected system. It is by far the cheapest option. Or, you can install a grid-connected system with battery backup. If you don't have access to the grid, you can install an off-grid system. All of these systems can combine two or more sources of electricity, creating hybrid systems.

When I consult with clients who are thinking of building passive solar/solar electric homes, I usually recommend a grid-connected system for those who live close to the utility grid. This configuration allows my clients to use the grid to "store" excess electricity, and it saves them a lot of money. Although they may encounter occasional power outages, in most locations in North America, these are rare and short-lived events.

Grid-connected systems with battery banks are suitable for those who want to stay connected to the grid but also want to protect themselves from occasional blackouts. They'll cost more, but they provide peace of mind and real security.

Off-grid systems are the system of choice for customers in remote rural locations. When building a new home in a rural location, grid connection can be pricey. The money you save by *not* connecting could pay for your PV system. Do remember, however, that off-grid homes need to be superefficient!

CHAPTER 6

Understanding Inverters

The inverter is an indispensable component of virtually all solar electric systems. It works long hours converting the DC electricity generated by a PV array into AC electricity, the type used in our homes and businesses. In battery-based systems, inverters contain circuitry to perform a number of additional useful functions.

In this chapter, we'll examine the role inverters play in renewable energy systems and then take a peek inside this remarkable device to see how it operates. We'll discuss all three types of inverters and discuss the features you should look for when purchasing one. But the first question to answer is this: Do you need an inverter?

Do You Need an Inverter?

Although this may seem like a ridiculous question, it's not. Some solar applications operate solely on DC power and, as a result, don't require inverters. Included in this category are small solar electric systems that power a few DC circuits in rustic out-of-the-way cabins and cottages, recreational vehicles, or sailboats. It also includes direct water-pumping systems that produce DC electricity to power DC water pumps, for example, for livestock. My DC solar pond aerator and the DC fans in my Chinese greenhouse are powered directly by PV modules.

Virtually all other renewable energy systems require an inverter. The type of inverter one needs depends on the type of system.

Types of Inverters

Inverters come in three basic types: those designed for batteryless grid-connected systems, those designed solely for off-grid systems, and those made for grid-connected systems with battery backup. Table 6.1 lists alternative names for inverters and PV systems that you may encounter.

Table 6.1. PV System and Inverter Terminology

Common terminology used in this book (referring to systems and inverters)	UL 1741 (Refers to inverter)	NEC 690- 2 (Referring to system)
Grid-connected	**Utility- interactive:** Operates in parallel with the utility grid.	**Interactive:** A solar photovoltaic system that operates in parallel with and may deliver power to an electrical production and distribution network.
Grid-connected with battery backup	**Multimode:** Operates as both or either stand-alone or utility interactive.	(Interactive, not separately defined)
Off-grid	**Stand-alone:** Operates independent of utility grid.	**Stand-alone:** A solar photovoltaic system that supplies power independently of an electrical production and distribution network.
		Hybrid: A system composed of multiple power sources. These power sources may include photovoltaic, wind, microhydro generators, engine-driven generators, and others, but do not include electrical production and distribution network systems. Energy storage systems, such as batteries, do not constitute a power source for the purpose of this definition.

Grid-Connected Inverters

Batteryless grid-connected PV systems require inverters that operate in sync with the utility grid. Two types are available: string inverters and microinverters. Both convert DC electricity produced by solar modules into AC electricity. Once this conversion takes place, electricity then flows from the inverter to the main service panel. It is then fed into active branch circuits, powering refrigerators, computers, stereo equipment, lights, and the like.

As noted in the previous chapter, the output of batteryless grid-connected inverters perfectly matches the voltage and frequency of electricity flowing through the electrical wires in our communities. Because of this, they can safely backfeed surpluses onto the utility "grid"

To produce the electricity that matches the grid voltage and frequency, these inverters continuously monitor electricity flowing through the utility lines. If an inverter detects a slight increase or decrease in voltage, it adjusts its output. If, for example, the voltage drops from 243 volts to 242.8 volts, the inverter automatically adjusts its output to conform precisely with the grid.

Grid-compatible inverters are equipped with *anti-islanding protection.* This consists of hardware and software that disconnects the inverter from the grid if the grid goes down. That way, a solar system doesn't become an island of electricity in an otherwise dead grid. This feature protects utility workers from electrical shock or, more likely, electrocution (fatal shock).

Grid-tie inverters also shut down if there's an increase or decrease in the frequency or voltage of grid power outside the inverter's acceptable limits. These conditions are referred to as *over/under voltage* or *over/under frequency.* If either the voltage or the frequency varies from the settings programmed into the inverter, it turns off—for example, if the frequency drops from 60 to 59.8 cps.

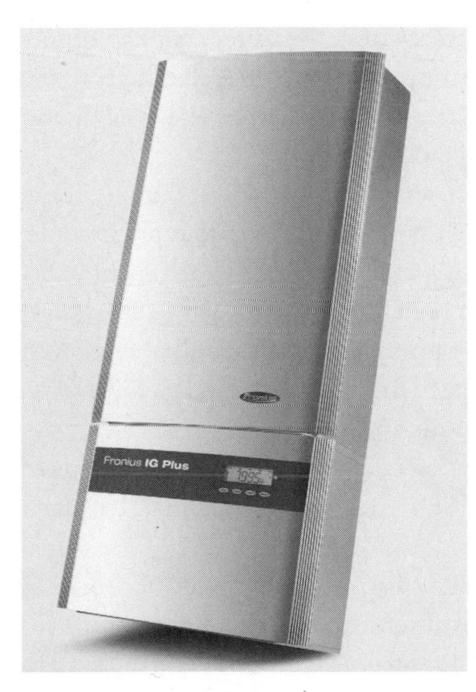

Figure 6.1. *Fronius Utility-Intertie Inverter. This sleek, quiet inverter from Fronius is suitable for grid connected systems without batteries. Fronius's inverters are produced at a production facility in Sattledt, Austria, which started production in early 2007. The facility receives 75% of its electricity from a large, roof-mounted PV system and 80% of its heat from a biomass heating system.*
Credit: Fronius.

Underfrequency and undervoltage can occur on hot, sunny days when numerous homeowners and businesses are running air conditioners. This condition is commonly referred to as a "brownout." The inverter remains off until the grid power returns within the acceptable range.

One fact that many who are contemplating a grid-tied solar electric system frequently have difficulty accepting is that when the grid goes down or experiences over/under voltage or over/under frequency, their system goes down. Not only does the inverter stop backfeeding surplus onto the grid, the inverter shuts off, terminating the flow of electricity to the main panel and to active branch circuits.

While that may seem ridiculous, remember the inverter must shut down to protect line workers, as just noted. However, because it needs a grid connection to determine the frequency and voltage of the AC electricity it produces, it is forced to terminate operations at least temporarily. Without the connection, the inverter can't operate. It's just wired that way.

To avoid losing power when the grid goes down, you can install a grid-connected system with battery backup. These systems allow homes and businesses to continue to operate when the utility grid has failed. Although inverters in such systems disconnect from the utility during outages, they continue to draw electricity from the battery bank to supply active circuits wired into a critical loads panel. It provides electricity to the most important (critical) loads. This transition takes place so quickly and smoothly that you won't even know it occurred.

Another option is SMA's line of inverters, Secure Power Supply. These grid-tied inverters can be ordered with a Secure Power Supply module. They terminate the flow of electricity to the grid and to the main panel during power outages, but continue to produce power (up to 1500 watts) so long as the Sun shines. Power is sent to an electrical outlet wired into the inverter. An extension cord plugged into this dedicated circuit can be run to critical loads such as refrigerators and freezers (Figure 6.2).

The only problem with this inverter is that it only supplies power during daylight hours. When the Sun goes downs, the inverter shuts down until sunrise.

Grid-connected inverters have LCD displays that provide information on a variety of parameters. The most useful are the AC output of the inverter in watts and the number of kilowatt-hours produced by the system

Figure 6.2. *This series of inverters ranging from 3 to 7.7 kW comes with an option to add a Secure Power Supply, software and hardware that allows the inverter to continue to produce electricity when the grid goes down. It will send that electricity to an electrical outlet (not shown) attached to the inverter, but not the grid. Total output of the inverter under such conditions is limited to 1,500 watts.* Credit: SMA

during a day. They also provide data on the total number of kilowatt-hours produced by a system since the inverter was installed.

Of lesser importance are the voltage and current (amperage) of incoming DC electricity and the voltage of the AC electricity backfeeding onto a grid. (One model, the Fronius inverter seen in Figure 6.1, keeps tab of the dollars saved and pounds of carbon dioxide emissions that you avoided by generating solar electricity.)

Another key feature found in many modern inverters—both string inverters and microinverters—is a function known as *maximum power point tracking* (MPPT). For reasons beyond the scope of this book, it ensures that a solar array always produces its maximum power output (in watts) under all temperature and irradiance conditions.

Grid-connected inverters operate with a fairly wide DC input range—typically between 150 to 500 volts DC. This permits them to use a wider range of modules and system configurations. As noted in previous chapters, higher-voltage wiring allows us to transmit electricity more efficiently. It also permits the use of slightly smaller and less expensive copper wires. With the cost of copper skyrocketing, savings incurred by using smaller wire size can be substantial.

Another advantage of high-voltage wiring of arrays is that it allows installers to place arrays farther from the inverter than low-voltage arrays. My company has installed PV arrays from 100 to 1,000 feet from homes. (Really long runs, like our 1,000-footers, however, require much larger wires to reduce line loss.)

Numerous companies produce utility-intertie string inverters. They can often be wired into three or four series strings of modules. These companies include OutBack Power, Fronius, SMA, Schneider Electric (formerly Xantrex, which was formerly Trace), and SolarEdge. A number of companies also produce microinverters, module-scale inverters, including Enphase, Siemens (they make microinverters for Enphase), SMA, SolarEdge, and ABB.

Inverters have long been the weak point in grid-tied solar electric systems. Fortunately, manufacturers have made changes to enhance performance. Today, string inverters typically come with 5- or 10-year warranties. Microinverters, first introduced in 1993, though only more recently becoming popular, come with a 25-year warranty. To study inverter options in more detail, be sure to check out *Home Power* magazine. *Home Power* regularly publishes articles about the inverters available in North America.

Figure 6.3a and b. *Inverter Installed Outside and Inside. Many installers place inverters that service large arrays near the array where they are exposed to the elements (a). I prefer to install inverters indoors like the one shown here (b). It provides added protection and ensures a longer lifespan. If you do install outdoors, be sure the inverter is shaded from the Sun throughout the year. The installer who put in the array shown on the left did shade the inverter, but only for six months of the year. Remember, the north side of an array will be illuminated early and late in the day from the spring equinox to the fall equinox. Be sure to protect against this sunlight, too.* Credit: Dan Chiras.

6.3b

Another product that you will encounter in your research is the *power optimizer.* These devices look a lot like microinverters and are wired directly to solar modules. They contain MPPT technology that increases the output of each module before it delivers DC power to a specially designed string inverter. This maximizes the efficiency of the solar array. According to SolarEdge, optimizers last much longer than microinverters, although they come with the same 25-year warranty. I've never used them but know installers who absolutely love them. Their common praise is that they are more reliable than microinverters.

As a final note, inverters should be installed *inside,* preferably in cool, dust-free environments. Inverters are rated for outdoor installation and can tolerate low as well as high temperatures and moisture, but most installers place their inverters outside for cost-savings, as shown in Figure 6.3a. I always install them in more hospitable environments to ensure a longer productive life. I can't help but think that exposure to extremes in weather as well as rain and snow will shorten an inverter's lifespan. As you shall soon see, inverters operate more efficiently when kept cool.

Off-Grid Inverters

Off-grid solar systems were once the mainstay of the solar energy business. Today, however, they constitute a tiny fraction of PV systems installations. However, off-grid systems are very popular among people seeking self-sufficiency, especially in rural areas. They're the only option

Figure 6.4. *Battery-Based Inverter. This inverter manufactured by Schneider Electric is designed for use on the grid with batteries as well as off grid. For ease of wiring, you can purchase this or similar inverters from other manufacturers pre-wired. That is, they come with all the breakers and internal wiring on the DC and AC side. This makes installation a snap, and saves a lot of brain damage, as wiring these systems can be quite complex.* Credit: Schneider Electric.

for those in really remote areas. Off-grid systems require battery-based inverters (Figure 6.4).

Like grid-connected inverters, off-grid inverters convert DC electricity into AC and boost voltage to 120 or 240 volts. Bear in mind, however, that many off-grid homes are wired for 120-volt AC. That's because it is impractical to install energy-guzzling 240-volt electric clothes dryers, electric stoves, electric water heaters, electric space heat, or heat pumps. These devices consume way too much electricity for the limited battery storage of off-grid systems.

Battery Charging

All inverters used in off-grid systems contain battery chargers that, as noted in the previous chapter, charge batteries from an external source—notably, a gen-set. Battery chargers convert AC from generators into DC which is then fed into the battery bank. Because off-grid inverters contain battery chargers, they are referred to as *inverter/chargers.*

Gen-sets are used in off-grid systems to restore battery charge after periods of deep discharge—for example, multiple cloudy days. Doing so reduces damage to the lead plates and increases battery life. Battery chargers are also called into duty during a battery maintenance operation called *equalization.* They are also critical to increasing the life of solar batteries as discussed in Chapter 7.

ABNORMAL VOLTAGE PROTECTION

High-quality battery-based inverters also contain programmable high- and low-voltage disconnect switches. These features protect various components of a system.

The low-voltage disconnect (LVD), for example, turns the inverter off when battery voltage drops below a certain point, indicating that the batteries are deeply discharged. The inverter stays off until the batteries are recharged. Thus, LVDs protect batteries from very deep discharges.

To avoid the hassle of having to manually start a generator when batteries are deep discharged, some inverters contain a sensor and switch that automatically activates a backup generator. When low battery voltage is detected, the inverter sends a signal to start the generator. (This only works if the generator has a remote start capability.) The generator then recharges the batteries. As you might suspect, more sophisticated auto-start generators cost more than the standard pull-cord type.

HIGH-VOLTAGE PROTECTION

Inverters in battery-based systems also contain a high-voltage shutoff feature. This sensor/switch terminates the flow of electricity from the gen-set when the battery voltage is extremely high. High-voltage protection prevents overcharging, which, like deep discharging, can severely damage the lead plates of batteries. This feature also protects the inverter from excessive battery voltage.

Inverters for Grid-Tied Systems with Battery Backup: Multimode Inverters

Although you may not want to go off grid, you may want to consider installing batteries for backup power. Grid-connected systems with battery backup are popular among homeowners who can't afford to be without electricity for a moment or those who experience frequent power outages lasting more than a few hours.

Grid-tied PV systems with battery backup require a special type of inverter: a battery- and grid-compatible inverter. They're commonly referred to as *multifunction* or *multimode inverters.*

Multifunction inverters contain features of grid-connected and off-grid inverters. Like a grid-connected inverter, they contain anti-islanding protection. Like off-grid inverters, multimode inverters contain battery chargers and high- and low-voltage disconnects.

Grid-tied systems with battery backup are not the most efficient PV systems. That's because a portion of the electricity generated in such systems is used to keep the batteries topped off. (The batteries are trickle-charged to offset self-discharge.) This may only require a few percent of a PV array's output, but over time, a few percent adds up. Remember also that the amount of energy required to trickle-charge batteries will increase as batteries age. That's because, as batteries age, they become less efficient, so more power is consumed to maintain battery voltage.

Before we move on, let me add a side note to those who are going off-grid. I'd strongly recommend that you consider installing a multifunction inverter just in case you decide at a later date to connect to the grid. Installing a multifunction inverter will save you a lot of money.

Retrofitting a Grid-Tied System with Batteries

As noted in Chapter 5, many states have passed laws that allow utilities to charge PV-powered customers more for electricity they purchase from the grid. This travesty has led many homeowners to contemplate cutting their ties with the utility and become their own power company—that is, going off-grid.

While going off grid is alluring, it can be challenging. One reason is that battery banks store so little electricity and most people consume huge amounts of electricity on a daily basis—more than a typical solar battery bank can store. If you are an energy-guzzler, switching to an off-grid system will require a major life change.

If you just want to convert a batteryless grid-tied system to a grid-tied system with battery backup, you've got two options. One is called *DC coupling*; the other is *AC coupling.*

DC coupling is a strategy employed by energy misers—people who are willing to go off grid and live that much more energy-efficient lifestyle. To do so, a homeowner must replace his or her grid-tied inverter with an off-grid inverter and install a battery bank (Figure 6.5). (He or she may have to rewire the solar array, too.) If one had initially installed microinverters, they'll need to be retired, and be replaced by an off-grid inverter and battery bank. A charge controller will also need to be installed and, of course, the homeowner will need to disconnect from the utility. This will take a home completely off grid.

The second option is *AC coupling*. This strategy is a rather ingenious way of fooling your grid-tied inverter to continue operating when the grid goes down. It's a bit complicated, however. As illustrated in Figure 6.6, in

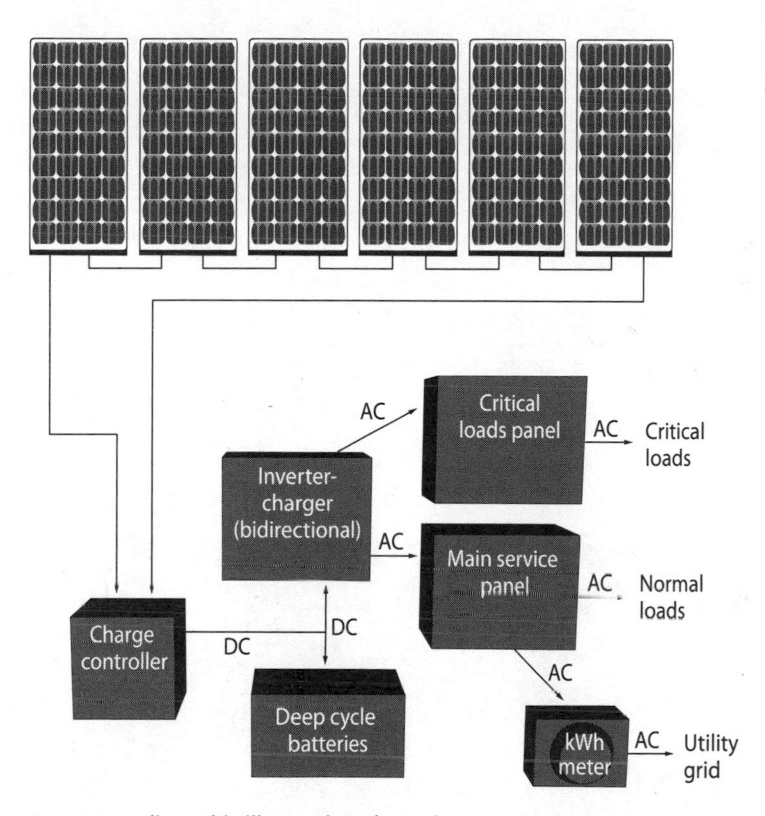

Figure 6.5. *DC Coupling. This illustration shows how a grid-tied system can be modified into a grid-tied system with battery backup. This makeover requires a homeowner to scrap his or her grid-tied inverter and replace it with a brand new multimode inverter.* Credit: Forrest Chiras.

this strategy you must install a battery bank and new multimode inverter, labeled "inverter-charger." Your existing grid-tied inverter is then wired into a newly installed subpanel. It contains all your critical loads and is appropriately called a *critical loads panel.* The inverter-charger is wired into the main service panel that connects to the grid. So how does this work?

When the grid is up and running, the PV array produces DC electricity that is fed into the grid-tied inverter (labeled "batteryless inverter"). The AC electricity it produces is fed into the critical loads panel where it powers active loads supplied by that panel. From the subpanel, unused AC electricity travels to the new inverter you installed, the inverter-charger. From the inverter-charger, the AC electricity flows to the battery bank to keep it full and to the main panel where it supplies the rest of

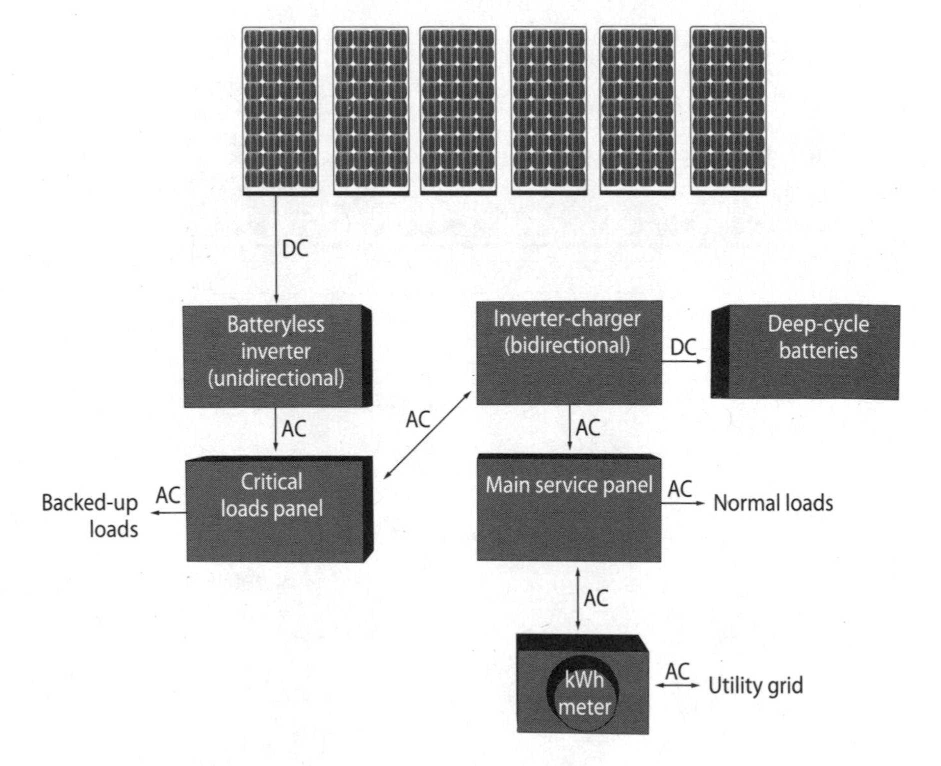

Figure 6.6. *AC Coupling. This illustration shows another way a grid-tied system can be modified into a grid-tied system with battery backup. In this case, the homeowner keeps his or her batteryless grid-tied inverter, so long as it can be reprogrammed to work in this installation. He or she adds a multimode inverter, battery bank, and a critical loads panel.* Credit: Forrest Chiras.

your active loads (noncritical loads). Surpluses, if any, are backfed onto the grid.

If the power goes out, however, the inverter-charger severs its connection with the grid, as required by Code. It does so via an automatic shutoff switch. The inverter-charger begins to draw electricity from the batteries. It feeds this electricity into the critical loads panel. The grid-tied inverter, in turn, senses normal voltage and continues to operate, converting DC power from the array into AC power. In essence, then, the inverter-charger tricks the grid-tied inverter into thinking that everything's okay. Because it and the inverter-charger only feed electricity into the critical loads panel, all is well. Ingenious, eh?

Installing systems such as this can be tricky, so consult a professional who has experience in such matters. This configuration may void warranties on grid-tied inverters. Also note that only specially programmed inverter-chargers can be used. You may have to swap out your existing batteryless grid-tied inverter for one that can be programmed to work in this wacky new system.

Buying an Inverter

Inverters come in many shapes, sizes, and prices. The smallest inverters, referred to as *pocket inverters,* range from 50 to 200 watts. They are ideal for supplying small loads such as VCRs, computers, radios, televisions, and the like. Most home and small businesses, however, require inverters in the 2,500 to 12,000-watt range. Which inverter should you select?

Type of Inverter

If you are going to have a system installed, your installer will specify the inverter. Installers usually purchase solar equipment from a local distributor. They, in turn, typically carry inverters from one or two manufacturers. Your installer will most likely install his or her favorite.

If you are going to install your own system, you'll need to select a reliable inverter that matches your system. Designing and installing a PV system can be quite challenging. There is a lot to know. It's a task best undertaken by those with a good working knowledge of electrical wiring and solar electric systems. A few hands-on PV installation workshops are extremely helpful.

Before we begin to explore things to look for when shopping for an inverter, note that you can purchase solar kits online from a number of companies like Northern Arizona Wind and Sun. Kits include all the components you'll need to install a system and should include wiring diagrams. While a kit is a great option, be sure to check out the equipment, especially the inverter. Find out where it is manufactured and the details of its warranty. Be sure to check out where equipment can be serviced. You don't want to be stuck with an inverter that can only be serviced in China or Israel. It is best to buy kits that come with equipment from major solar manufacturers in your country with nearby service centers.

The first consideration when shopping for an inverter is the type of system you are installing. If you are installing a grid-connected system without batteries, you'll need an inverter designed for such systems. If you are installing an off-grid system, you'll need an off-grid inverter/charger. If you are installing a grid-connected system with battery backup, you must purchase an inverter that is both grid- and battery-compatible—a multimode inverter. As noted earlier, multimode inverters may also be a good choice for off-grid systems, just in case you decide to connect to the utility at a later date.

You can get help choosing an inverter from a local or an online supplier; however, the more you know, the more their advice will make sense. Let's start with grid-tied inverters.

Choosing a Grid-Tied Inverter

Shopping for a grid-tied inverter is relatively easy. If you know the size of your solar array, you need to buy an inverter that matches it. For example, if you find that you need a 5-kW array to meet your electrical needs, you will need a 5-kW batteryless grid-tied inverter. (This inverter will service systems from 4.8 to 5.2 kW.)

After that, most people shop for efficiency. Modern grid-tied inverters tend to be around 96% efficient. Add optimizers and you'll boost the efficiency a bit more. Install microinverters rather than a string inverter and you'll also enjoy a slightly higher efficiency.

Be sure the inverter is UL 1741 listed. This ensures the utility that your inverter has anti-islanding capability. Also be sure to check out the warranty. Most inverters come with a 5-year warranty, although 10-year warranties are starting to become more common. In some instances, you may be able to purchase an extended warranty for a string inverter. I'd

PV Array Voltage, String Size, and Choosing the Correct Inverter

To help installers determine how many modules they can wire in a string, virtually all inverter manufacturers provide online calculators. To calculate string size, simply enter the temperature conditions of the site (they'll ask for low-temperature data) and the type of module you are going to install. The online calculator provides the maximum, minimum, and ideal number of modules in a series string (string length) for each of a manufacturer's inverters.

Temperature is important when determining the size of series strings in a PV array because the output of a PV array changes with temperature. Low temperatures, for example, increase the voltage of an array. If an array has not been sized carefully, the voltage of the incoming DC electricity could exceed the rated capacity of the inverter, causing serious damage.

High temperatures, in contrast, reduce the output of an array. If the voltage of a PV array falls below the rated input of a grid-tie inverter, a PV system will shut down and will remain off until the array cools down and the voltage increases. This, of course, reduces the efficiency of a system.

strongly recommend it. Microinverters and optimizers come with 25-year warranties. I only install microinverters.

It is important to check the output voltage of the inverter. Homes are typically wired at 240 volts, so you'll probably need a 240-volt inverter. Commercial buildings are typically wired at 208, 480, or 600 volts. Be sure your inverter matches the voltage of your application.

When shopping for an inverter, be sure to study the DC voltage input range so you can determine the length (number of modules) of series strings you can wire into that inverter. As you may recall, the number of modules affects the voltage of each series string. Most string inverters can be wired to series strings of 10 to 12 modules. For those who want to explore this topic in more detail, see the accompanying sidebar.

Choosing a Battery-Based Inverter

If you are going off-grid or installing a grid-tied system with battery backup, you'll find that choosing an inverter requires a little more know-how. In this section, I will discuss the key factors that go into choosing a battery-based inverter, including (1) input voltage, (2) waveform, (3) continuous output, (4) surge capacity, and (5) efficiency. I'll also discuss battery chargers, noise, and a few other factors to consider.

Inverter Input Voltage

When shopping for a battery-based inverter, you'll need to select one with an input voltage that corresponds with the voltage of your battery bank. (That's because in these systems batteries feed DC electricity to inverters.) It's not as complicated as it sounds.

Battery banks in solar electric systems are typically wired at either 12-, 24-, or 48-volts. Twelve-volt systems are common in cabins and summer cottages. Residential systems are either 24- or 48-volt systems, most commonly 48 volts, because they're the most efficient. Be sure the input voltage of the inverter you choose matches your battery bank's voltage.

Waveform: Modified Square Wave vs. Sine Wave

The next inverter selection criterion you must consider is the output waveform. What's that?

Waveform is a rather technical topic. It is a measure pertaining to the voltage of alternating current electricity. Unfortunately, it is beyond the scope of this book.

Remember this, though: battery-based inverters for off-grid systems inverters are available in two types: *modified square wave* (often called modified sine wave) and *sine wave*. Because the wave form of electricity on the electrical grid is sine wave, all grid-connected inverters produce sine wave AC electricity. That way, their output matches utility power. Off-grid inverters can be either sine wave or modified square wave. Multimode inverters for grid-tied systems with battery backup must be sine wave.

I strongly recommend sine wave inverters no matter what your application, even if going off-grid. Modified square wave electricity is a crude approximation of grid power. It works well in many appliances such as refrigerators, older washing machines, and power tools. It also works pretty well in most electrical devices, including TVs, lights, stereos, computers, and inkjet printers. Although all these devices can operate on this lower-quality waveform, they run less efficiently, producing more heat and less work—light or water pumped, etc.—for a given input.

Bigger problems arise, however, when modified square wave is fed into sensitive electronic circuitry such as microprocessor-controlled front-loading washing machines, appliances with digital clocks, chargers for various cordless tool, copiers, computers, and laser printers. These devices require sine wave electricity. Without it, you're sunk. In my off-grid home in Colorado, for example, I found that my energy-efficient front-loading Frigidaire Gallery washing machine would not run on the modified square wave electricity produced by my inverter. The microprocessor that controls this washing machine—and other similar models—simply can't operate on this inferior form of electricity. After I replaced the inverter with a sine wave inverter, I had no troubles whatsoever.

Making matters worse, I found that some electronic equipment, such as TVs and stereos, gave off an annoying high-pitched hum when operating on modified square wave electricity. Modified square wave electricity may also produce annoying lines on TV sets and can even damage sensitive electronic equipment.

When operated on modified square wave electricity, microwave ovens cook more slowly. Equipment and appliances also run warmer and might last fewer years. Computers and other digital devices operate with more errors and crashes. The only time I have burned out a computer was when I was powering my off-grid home with a modified square wave inverter—and I've been working on computers since 1980. In addition to these problems, digital clocks don't maintain their settings as well when operating on modified square wave electricity. Moreover, motors don't always operate at their intended speeds. So why do manufacturers produce modified square wave inverters?

The most important reason is cost. Modified square wave inverters are much cheaper than sine wave inverters. You will most likely pay 30% to 50% less for one.

Another reason for the continued production of modified square wave is that they are hardy beasts. They work hard for many years with very little, if any, maintenance. (Their durability may be related to their simplicity: they are electronically less complex than sine wave inverters.) Regardless, I recommend that you purchase a sine wave battery-based inverter for off-grid systems. Their output is well suited for use in modern homes with their array of sensitive electronic equipment. SMA, Schneider Electric, and OutBack all produce excellent battery-based sine wave inverters.

Continuous Output

Continuous output is a measure of the power a battery-based inverter can produce on a continuous basis—provided there's enough energy available in the system. The power output of an inverter is measured in watts, although some inverter spec sheets also list continuous output in amps (to convert watts to amps, use the formula watts = amps × volts). Schneider Electric's (formerly Xantrex) sine wave inverter, model SW2524, for instance, is capable of producing 2,500 watts of continuous power. This inverter can power a microwave using 1,000 watts, an electric hair dryer using 1,200 watts, and several smaller loads simultaneously without a hitch. (By the way, the "25" in the model number indicates the unit's continuous power output; it stands for 2,500 watts. The "24" indicates that this model is designed for a 24-volt PV system.) The spec sheet on this inverter lists the continuous output as 21 amps.

OutBack's sine wave inverter VFX3524 produces 3,500 watts of continuous power and is designed for use in 24-volt systems. Off-grid homes can easily get by on a 3,000- to 4,000-watt inverter.

To determine how much continuous output you'll need, add up the wattages of the common appliances you think will be operating at once. Be reasonable, though. Typically, only two or three large loads operate simultaneously. A washer and well pump may be operating, for example, when a sump pump is running. If you are planning to operate a shop next to your home on the same inverter, you will need an inverter with a higher continuous output or you will need to install a generator.

Surge Capacity

Electrical devices with motors, such as vacuum cleaners, refrigerators, washing machines, and power tools, require a surge of power to start up. It is a momentary spike in amperage required to get the motor running. The spike typically lasts only a fraction of a second. Even so, if an inverter doesn't provide a sufficient amount of power, the power tool or appliance won't start! Not starting is not just inconvenient. A motor that's not receiving enough power to start will draw excessive current and overheat very quickly. Unless they are protected with a thermal cutout, they may burn out.

When shopping for an inverter, be sure to check out surge capacities. All quality solar inverters are designed to permit a large surge of power over a short period, usually about five seconds. Surge capacity or surge power is listed on spec sheets in either watts and/or amps.

EFFICIENCY

Another factor to consider when shopping for inverters is their efficiency. Efficiency is calculated by dividing the energy coming out of an inverter by the energy going in and then multiplying by 100 to get a percentage.

Most battery-based inverters models have efficiencies in the mid to low 90s. Although there's nothing you can do about it, you should note that the efficiency of inverters varies with load—how much power is being used. Generally, an inverter achieves its highest efficiency once output reaches 20% to 30% of its rated capacity. A 4,000-watt inverter, for instance, will be most efficient at outputs above 800 to 1,200 watts. At lower outputs, efficiency is dramatically reduced.

COOLING AN INVERTER

Because they're not 100% efficient, inverters produce waste heat internally. A 4,000-watt inverter running at rated output at 90% efficiency produces 400 watts of heat internally. (That's equivalent to the heat produced by four 100-watt incandescent light bulbs.) The inverter must get rid of this heat to avoid damage.

Inverters rely on cooling fins to passively rid themselves of excess heat. Fins increase the surface area from which heat can escape. Some inverters also come with fans to provide active cooling. Fans blow air over internal components, stripping off heat. If the inverter is in a hot environment, however, it may be difficult for it to dissipate heat quickly enough to maintain a safe temperature.

Unbeknownst to many, inverters are programmed to power down (produce less electricity) when their internal temperature rises above a certain point. Translated, that means if you need a lot of power, and the internal temperature of your inverter is high, you won't get it. A Schneider Electric's SW series inverter, for example, produces 100% of its continuous power at 77°F (25°C), but drops to 60% at 117.5°F (47.5°C).

The implications of this are many. First, as I've pointed out before, inverters should be installed in relatively cool locations. If the inverter is mounted outside, make sure it is shaded. Second, inverters should be installed so that air can move freely around them. Don't box in an inverter to block noise (though not all inverters are noisy).

Battery Charger

Battery chargers are standard in battery-based inverters. As you may recall, a battery charger will allow you to charge your DC battery bank using AC from the utility (for grid-connected systems with battery backup) or an AC generator (for either off-grid systems or grid-connected systems with battery backup).

Noise and Other Considerations

Battery-based inverters are typically installed inside not so much for weather protection, although that's important, but so that they can be located close to the batteries, which do best in a nice, comfortable temperature. Inverters are also located close to batteries to reduce line loss and cost. The wires from battery banks carry either 12-, 24-, or 48-volt DC. To reduce line loss in such runs, it is best to minimize the length of wire runs. Code requires the use of very large wires as well, which are expensive; so, the shorter the run, the less it will cost.

If you are planning on installing an inverter inside your home or business, be sure to check out the noise it produces. Inquire about this upfront, or, better yet, ask to listen to the model in operation to be sure it's quiet. Don't take a manufacturer's word for it. My first inverter was described by the manufacturer as "quiet," but I found out that quiet meant compared to a Boeing 707. It emitted an annoyingly loud buzz heard throughout my house. The first six months after I moved into my off-grid, rammed earth tire home, the inverter's buzz drove me nuts, but I grew used to it. Fortunately, battery-based inverters on the market today are pretty quiet.

Some folks are also concerned about the potential health effects of extremely low-frequency electromagnetic waves emitted by inverters and electrical wires. If you are concerned about this, install your inverter in a place away from people. Avoid locations in which people will be spending a lot of time—for example, don't install the inverter on the other side of a wall from your bedroom or office.

Your checklist of features to consider when purchasing an inverter should include ease of programming. Find out in advance how easy it is to change settings, and don't rely on the opinions of salespeople or renewable energy geeks that can recite pi to the 20th decimal place. Ask friends or dealers/installers for their opinions, but also ask them to show you. You may even want to spend some time with the manual (often available online) to see if it makes sense *before* you buy an inverter.

Another feature to look for in a battery-based inverter is power consumption under search mode. What's that?

The search mode is an operation that allows a battery-based inverter to go to sleep, that is, shut down almost entirely in the absence of active loads. The search mode saves energy.

Although the inverter is sleeping when it is in search mode, it's sleeping with one eye slightly open. That is, it's on the alert should someone switch on a light or an appliance. It's able to do this by sending out tiny pulses of electricity approximately every second. They are sent through the branch circuits like scouts seeking an active load. When an appliance or light is turned on, the inverter senses the load and quickly snaps into action, powering up and feeding AC electricity.

The search mode is handy in houses in which phantom loads have been eliminated. As you may recall, a phantom load is a device that continues to draw a small electrical current when off. You may also recall that phantom loads, on average, account for about 5% of a home's annual electrical consumption. (In some homes, however, they can be as high as 10%.)

Eliminating phantom loads saves a small amount of electrical energy in an ordinary home. In a home powered by renewable energy, however, it saves even more because supplying phantom loads 24 hours a day requires an active inverter. An inverter may consume about 30 watts when operating at low capacity. Servicing a 12-watt phantom load, therefore, requires an additional 30-watt investment in energy in the inverter—24 hours a day.

Energy savings created by eliminating phantom loads and continuous inverter operation does have its downsides. For example, automatic garage door openers may have to be turned off for the inverter to go to into the search mode. In my off-grid home in Colorado, I turned off my garage circuit at the breaker box or main service panel when I'd leave home. It was a pain in the neck, however, because when I'd arrive back home, I had to get out of the car, walk into the house, then switch it back on. (Ah, the things we do to live sustainably!)

Another problem occurs with electronic devices that require tiny amounts of electricity to operate, like cell phone chargers. When left plugged in, cell phone chargers may draw enough power to cause an inverter to turn on. Once the inverter starts, however, the device doesn't draw enough power to keep it going. As a result, the inverter switches on and off, *ad infinitum*. The phantom load drawn by my portable stereo caused this problem.

Because of these issues, many people simply turn the search mode off so that the inverter keeps running 24 hours a day. Bear in mind, that means you consume 30 watts × 24 hours per day, just to keep the thing going. That's 0.7 kilowatt-hours, which doesn't sound like much, but in a home that only consumes 5 kWh a day ... well, it's a lot.

Another option is to set the search mode sensitivity up, so it turns on at a higher wattage. However, most modern homes have at least one always-on load, for example, hard-wired smoke detectors or garage door openers that require the continuous operation of the inverter. Your inverter will be on 24 hours a day, 365 days a year, eating up a lot of energy.

Conclusion

If you are hiring an honest, experienced professional to install a PV system, you shouldn't have to worry about buying the right inverter. When installing your own system, be sure that your source takes the time to determine which inverter is right for you. Provide as much information about your needs as possible and ask lots of questions.

A good inverter is key to the success of a renewable energy system, so shop carefully. Size it appropriately. Be sure to consider future electrical needs. But don't forget that you can trim electrical consumption by installing energy-efficient electronic devices and appliances. Efficiency is always cheaper than adding more capacity. When shopping, select the features you want and buy the best inverter you can afford. Although modified square wave inverters work for most applications, it is best to purchase a sine wave inverter. It is a decision you will not regret if you plan to operate a computer or a TV.

CHAPTER 7

Batteries, Charge Controllers, and Gen-sets

If you're going to install a batteryless grid-tied system, I have good news for you. You can skip this chapter! If you are going to install a system with batteries—either an off-grid system or a grid-connected system with batteries—I've got some bad news. You've got a ton to learn. You'll need to know what your options are, how to size a battery bank, how to properly install batteries, and how to maintain them.

Even if you hire someone to install your system, you'll need to know a great deal about batteries so you can properly maintain them. If you don't take this job seriously, you'll destroy a lot of very costly batteries over the lifetime of your system. Putting a more positive spin on it: Taking care of your batteries will help you get the most from them and could save you a fortune over the long haul.

This chapter will help you develop a solid understanding of batteries and two additional components required in battery-based PV systems: charge controllers and generators.

Understanding Lead-Acid Batteries

Batteries are a mystery to many people. How they work, what makes one type different from another, and how they are damaged are topics that can boggle the mind. Fortunately, the batteries used in most off-grid renewable energy systems are pretty much the same: they're deep-cycle, flooded lead-acid batteries.

Flooded lead-acid batteries for renewable energy systems are the ultimate in rechargeable batteries. They can be charged and discharged (cycled) a thousand or more times before they wear out, provided you take good care of them.

Lead-acid batteries used in most renewable energy systems contain three cells—that is, three distinct 2-volt cells, or compartments. Inside the battery case, the individual cells are electrically connected (wired in series). As a result, they collectively produce 6-volt electricity.

Inside each cell in a flooded lead-acid battery is a series of thick, parallel lead plates, as shown in Figure 7.1. Each cell is filled with battery fluid, hence the term "flooded." Battery fluid consists of 70% distilled water and 30% sulfuric acid. A partition wall separates each cell, so that fluid cannot flow from one cell to the next. The cells are enclosed in a heavy-duty plastic case.

As illustrated in Figure 7.1, two types of plates are found inside a battery: positive and negative. All the positive plates are wired together in series, as are all the negative plates. They are then connected to the positive and negative battery posts or terminals. Battery posts allow electricity to flow into and out of a battery.

As shown in Figure 7.1, the positive plates of lead-acid batteries are made from lead dioxide (PbO_2). The negative plates are made from pure lead. The sulfuric acid/water mix that fills the spaces between the plates is referred to as the *electrolyte.*

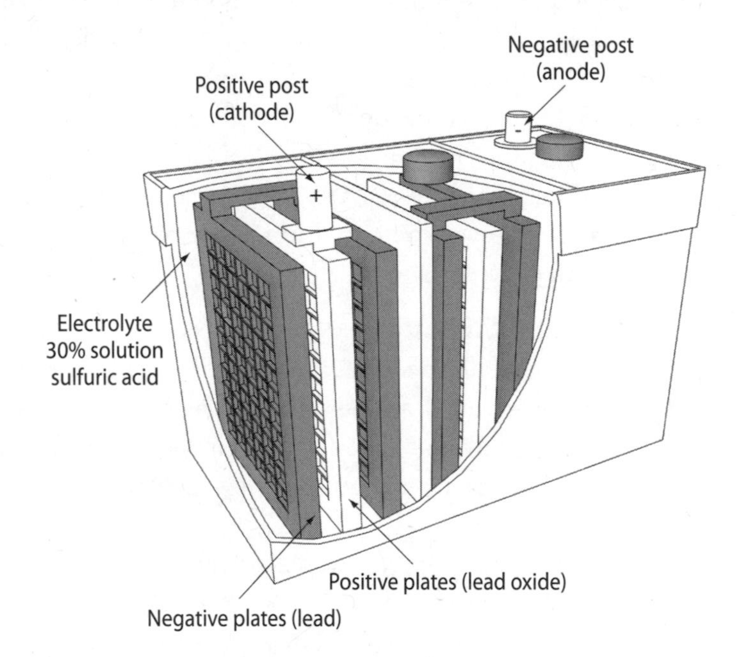

Figure 7.1. *Cross Section of Lead-Acid Battery.* Credit: Anil Rao.

How Lead-Acid Batteries Work

Like all other types of batteries, lead-acid batteries convert electrical energy into chemical energy when charging. When discharging (that is, giving *off* electricity), chemical energy is converted back into electricity. The details of the chemical reactions are shown in Figure 7.2, but before you take a look, be warned: it's pretty complicated chemistry. That said, you don't need to master this chemistry to figure out how to take care of your battery bank. You do need to understand a few key concepts, though, which are pretty easy to grasp. Let me walk you through the important points.

Note first that the top reaction depicts the chemistry occurring on the negative plates. The bottom reaction shows the chemical reaction at the positive plates. Note that both reactions show arrows pointing to the right. These reactions occur simultaneously when the batteries are discharging, releasing electricity.

As illustrated, when electricity is drawn from a lead-acid battery, sulfuric acid (HSO_4^-) reacts chemically with the lead in the negative and positive plates. As you can see in the top equation, this chemical reaction yields electrons ($2e^-$), tiny negatively charged subatomic particles. The electrons flow out of the battery via the negative terminal through the battery cable. This constitutes an electrical current. During this chemical reaction, you also will note that lead reacts with sulfate ions to form lead sulfate ($PBSO_4$). Thus, when batteries are discharging, tiny lead sulfate crystals form on the surface of the negative plates.

The second equation in Figure 7.2 shows the chemical reaction that takes place on the surface of the positive plates when a battery is discharging. It's a bit more complicated, but what's most important is that sulfuric acid also reacts with the lead (lead dioxide) of the positive plates, forming tiny lead sulfate crystals on them as well. Keep that in mind, because it's

$Pb(s) + HSO_4^-(aq) \rightarrow PbSO_4(s) + H^+(aq) + 2e^-$	negative plates
$PbO_2(s) + HSO_4^-(aq) + 3H^+(aq) + 2e^- \rightarrow PbSO_4(s) + 2H_2O(1)$	positive plates

Figure 7.2. *Chemical Reactions in a Lead-Acid Battery. The chemical reactions occurring at the positive and negative plates are shown here. Note that these reactions are reversible. Lead sulfate forms on both the positive and negative plates during discharge.* Credit: Anil Rao.

the formation of lead sulfate that reduces a battery's lifespan. Excessive buildup can destroy a battery. Your job as an off-grid solar system operator is to see to it that lead sulfate is quickly removed from the plates; you do this by reversing the chemical reaction—that is, charging the batteries as soon as possible after discharging.

When a battery is being charged by a solar array or a generator, the chemical reactions shown in Figure 7.2 take place in reverse. As a result, lead sulfate crystals on the surface of both the positive and negative plates break down. Sulfuric acid is regenerated. The plates are restored, although a very small number of lead sulfate crystals flake off, slowly but surely whittling away at the plates.

What is most important to remember is that these chemical reactions allow us to store energized electrons inside batteries and reclaim them when we need electricity.

A Word of Warning

Sulfuric acid is a *very* strong acid. In fact, it is one of the strongest acids known to science. In flooded lead-acid batteries, sulfuric acid is diluted to 30%. Although diluted, it is still to be treated with great respect—it can burn your skin and eyes and eat through clothing like a ravenous moth. Be sure to wear eye protection and rubber gloves when working with batteries, especially when filling them. It's a good idea to remove jewelry and wear a long-sleeved shirt you don't care about. It's bound to get a little acid on it; and the acid will produce tiny holes in the shirt that you won't notice until after you wash it.

Will Any Lead-Acid Battery Work?

Lead-acid batteries come in many varieties, each one designed for a specific application. Car batteries, for example, are designed and manufactured for use in cars, light trucks, and vans. Marine batteries are designed for boats; golf cart batteries for use in golf carts, and forklift batteries for forklifts.

For an off-grid solar system, you have three options: (1) deep-cycle flooded lead-acid batteries, like those made by Trojan, Rolls, and Deka

(Figure 7.3); (2) forklift batteries; or (3) golf cart batteries. Car batteries won't work. Their lead plates are much too thin to withstand deep cycling (deep discharges), which commonly occurs in renewable energy systems. Although lead sulfate crystals that form on the plates of a battery during deep discharge are removed when these batteries are recharged, some crystals fall off before recharge occurs. The thin plates of a car battery are therefore whittled away to nothing in a very short time. After 20 or so deep discharges, the batteries would be ruined—no longer able to accept a charge.

Renewable energy systems require deep-discharge lead-acid batteries with thick lead plates. The thickness of the lead plates allows them to withstand multiple deep discharges—so long as they're recharged soon afterward. Even though the plates lose lead over time, they are so thick that the small losses are insignificant. As a result, a deep-cycle battery can be deeply discharged hundreds, sometimes a few thousand times, over its lifetime.

For optimum long-term performance, however, deep-cycle batteries need to be recharged promptly after deep discharging. Don't ever forget this! With proper care, good, high-capacity deep-cycle batteries like the L16s most often used in PV systems can easily last seven to ten years. If you're really careful, you could get a dozen or more years of useful life out of a battery bank.

Figure 7.3. *Lead-Acid Batteries for RE System. Lead-acid batteries are commonly used in renewable energy systems. Be sure to purchase deep-cycle batteries; keep them in a warm, safe place; recharge quickly after deep discharges; and equalize them periodically.* Credit: Surrette Battery Company LTD.

Forklift batteries are high-capacity, deep-discharge batteries designed for a fairly long life, and they operate under demanding conditions. They can withstand 1,000 to 2,000 deep discharges—more than many other deep-cycle batteries used in battery-based renewable energy systems—so they work well in renewable energy systems. They are, however, rather heavy, bulky, and expensive. But, if you can acquire them at a decent price, you may want to use them.

Golf cart batteries may also work. Like forklift batteries, golf cart batteries are designed for deep discharge. But they typically cost a lot less than other heavier duty deep-cycle batteries. While the lower cost may be appealing, golf cart batteries don't last as long as the alternatives. They may last only five to seven years, if well cared for. Shorter lifespan means more frequent replacement. More frequent replacement means higher long-term costs and more hassle.

What About Used Batteries?

Another option for cost-conscious solar system owners is a used battery. Although used batteries can often be purchased at bargain prices, they're rarely worth it. Used batteries are often discarded because they've failed or have experienced a serious decline in function. As a buyer, you have no idea how well—or how poorly—they've been treated. Have they been deeply discharged many times? Have they been left in a state of deep discharge for long periods? Have they been filled with tap water rather than distilled water? Although there are exceptions, most people who've purchased used flooded lead-acid batteries have been disappointed.

When shopping for batteries for a renewable energy system, I recommend that you buy high-quality deep-cycle batteries. Although you might be able to save some money by purchasing cheaper alternatives, frequent replacement is time consuming and costly. Batteries are heavy, and it takes quite a lot of time to disconnect old batteries and rewire new ones. Bottom line: the longer a battery will last—because it's the right battery for the job and it's well made and well cared for—the better!

Sealed Batteries

Grid-connected systems with battery backup sometimes incorporate another type of lead-acid battery, known as *sealed lead-acid battery* or *captive-electrolyte battery* (Figure 7.4). Sealed lead-acid batteries are filled with electrolyte at the factory, charged, and then permanently sealed. This makes them easy to handle. They can be shipped without fear of spillage. They won't even leak if the battery casing is cracked, and they can be installed in any orientation—even on their sides. But most important, they never need to be watered.

Two types of sealed batteries are available: absorbed glass mat (AGM) batteries and gel cell batteries. In absorbed glass mat batteries, thin absorbent fiberglass mats are placed between the lead plates. The mat consists of a network of tiny pores that immobilize the battery acid. These tiny pockets also capture hydrogen and oxygen gases given off by the battery when it is charging. Unlike a flooded lead-acid battery, the gases can't escape. Instead, they recombine in the pockets, reforming water. That's why sealed AGM batteries never need watering.

In gel batteries, the sulfuric acid electrolyte is converted to a substance much like hardened Jell-O by the addition of a small amount of silica gel. The gel-like substance fills the spaces between the lead plates.

Sealed batteries are also known as "maintenance-free" batteries because fluid levels never need to be checked and because the batteries never need to be filled with water. They also never need to be (and should not be!) equalized, a process discussed later in this chapter. Eliminating routine maintenance saves a lot of time and energy. It makes sealed batteries a good choice for grid-connected systems with battery backup. In these

Figure 7.4. *Sealed Lead-Acid Battery. Sealed batteries like these 12-volt batteries never need to be watered. Nonetheless, they still require proper housing and temperature conditions and careful control of state of charge to ensure a long life. Notice the battery posts and nonremovable pressure-release caps.* Credit: Dan Chiras.

systems, batteries are rarely used and hence often forgotten and not regularly maintained. Sealed batteries are also ideal for off-grid systems in remote locations where routine maintenance is problematic—for example, in rarely occupied backwoods cabins or cottages.

Sealed batteries offer several additional advantages over flooded lead-acid batteries. They charge faster, and when charging they do not release explosive gases. As a result, there's no need to vent battery rooms or battery boxes. In addition, sealed batteries are much more tolerant of low temperatures. They can even handle occasional freezing, although this is never recommended. Sealed batteries self-discharge more slowly than flooded lead-acid batteries when not in use, too. (All batteries self-discharge when not in use.)

Unfortunately, sealed batteries are more expensive, store less electricity, and have a shorter lifespan than flooded lead-acid batteries. They also can't be rejuvenated (equalized) if left in a state of deep discharge for an extended period of time. During such periods, lead sulfate crystals on the plates grow in size. Large crystals then begin to reduce battery performance. As the crystals grow, batteries take progressively less charge. Because they can't be fully charged, they have less to give back. Over time, entire cells may die, substantially reducing the storage capacity of a battery bank.

Large crystals on the plates of flooded lead-acid batteries can be removed by a controlled overcharge, a procedure known as *equalization* Although equalization is safe in unsealed flooded lead-acid batteries, it results in pressure buildup inside a sealed battery. Pressure is vented through the pressure release valve on the sealed battery, causing electrolyte loss that could destroy the storage capacity of the battery.

Bottom line: while maintenance-free batteries may seem like a good idea, they are not suitable for many applications.

Lithium Ion Batteries

Lead-acid batteries are the weak link in off-grid and grid-tied systems with battery backup. They are costly, don't store much electricity, and require proper housing (not too hot, not too cold). They also require periodic maintenance and more-frequent-than-you-will-be-happy-with replacement. This has led manufacturers to search for better options. Interestingly, much of the research and development on new batteries is occurring in the electric car industry. Companies such as Tesla, Nissan, and Ford—all leaders in this field—are intensely interested in developing inexpensive, long-lasting, lightweight, and high-capacity batteries to extend the range of their electric vehicles. Today's electric cars, like my all-electric Nissan Leaf, are all powered by lithium ion batteries (Figure 7.5).

Tesla introduced a wall-mounted lithium ion battery called a Powerwall for use as a backup for solar homes (Figure 7.6). At this writing (February 2019), the company offers a 13.5-kWh battery. I'm often asked whether

Figure 7.5. *Dan with His Electric Car and Dog. The author's all-electric 2013 Nissan Leaf, which he powers entirely with solar and wind energy. Folks interested in buying a used all-electric car can get some killer deals on slightly used vehicles. This car costs over $35,000 new. He got it in January 2016 for under $9,000, and it was in perfect condition.* Credit: Linda Stuart.

Figure 7.6. *Tesla Powerwall Battery.* Credit: Tesla.

I recommend these batteries. Here's my thinking on the subject: A typical lead-acid battery bank in a 48-volt off-grid solar system would store over 50 kWh of electricity; however, to protect batteries and ensure long life, we don't like homeowners to use more than half of that, so a typical battery bank in an off-grid home really only stores 25 kWh of electricity. This lead-acid battery bank would cost about $7,500 plus installation, which is a significant figure. For an off-grid system, then, you'd need two Tesla Powerwalls, costing $14,500 plus installation. For grid-tied systems with battery backup, you'd need one Powerwall costing about $7,250 plus installation. A lead-acid battery bank consisting of eight 6-volt batteries would cost about $2,500 plus installation. Based on price, a conventional lead-acid battery bank would make more sense. But before you abandon the idea, consider some of the plusses.

Pros and Cons of Tesla Powerwall

The Powerwall is a nifty idea. The units come with a 10-year warranty. They can be mounted on the wall or on a floor, and they can be "stacked" (wired together) to increase storage capacity. Unfortunately, as just noted, they are quite costly. Installation isn't cheap either. According to Tesla, the installation will cost $1,000 to $2,000. Other sources note that installation costs are usually much higher, in the range of $5,000 to $8,000.

If you are thinking about taking your grid-tied home off grid, you'll find that this battery won't store as much electricity as is needed by most homes. Let me explain. Tesla's initial market for its battery was grid-tied customers who wanted to add storage so they could use the surplus electricity they produced during the day at night. But, with more and more utilities charging solar customers higher monthly fees and/or higher electrical costs, many solar homeowners are turning to the Powerwall to go completely off grid. The problem with both of these approaches is that most homes in the US and Canada consume at least 1,000 watts of continuous power—that's one kilowatt-hour of electricity per hour. If you're installing a Powerwall for backup power at night and are operating it from 5 p.m. to 8 a.m. in the winter, the battery will supply electricity for 14 hours. Although most people sleep at night and require less energy at night than during the day, the battery bank may not be sufficient to meet all needs for many households. If you prepare and consume your dinner after dark on an electric stove, have two or three flat-screen TVs operating at the same time, have several computers running at the same time, or wash or dry clothes in an electric clothes dryer, a Powerwall most likely won't be able to meet your demand. If the weather's cold and you are running a couple of space heaters, which use over 1,000 watts each, or an electric furnace, you'd be out of luck unless you install two or more Powerwalls.

Before you make the leap off grid, be sure to calculate how much energy storage you will need and how many batteries would be needed to make that happen (or ask the installer to run the calculations for you), so you can be sure you make a wise decision.

Wiring a Lead-Acid Battery Bank

Batteries are wired by installers to achieve a specific voltage. Small renewable energy systems—for example, those used to power RVs, boats, and cabins—are typically wired to produce 12-volt electricity. Many of the loads in these applications run entirely off 12-volt DC electricity. Systems in off-grid homes and businesses are typically wired to produce 24- or 48-volt DC electricity. In these systems, however, the low-voltage DC electricity is converted to AC electricity by an inverter. It boosts the voltage to the 120- and 240-volts AC commonly used in homes and businesses.

Sizing a Battery Bank

Properly sizing a battery bank is key to designing a reliable off-grid system. The principal goal when sizing a battery bank is to install a sufficient number of batteries to carry your household or business through periods when the sun or wind and sun (in hybrid systems) are not as readily available.

In sunnier locations, like Colorado, Wyoming, and Utah, battery banks in off-grid applications are often sized to meet the need for electricity for three days. Longer reserve periods—five days or more—may be required in other areas, for example, the cloudier Midwest, Northeast, and Southeastern states and much of Canada. As noted in Chapter 5, backup fossil-fuel generators are often included in off-grid systems. Backup generators can reduce the size of the battery bank.

For more details on wiring and sizing battery banks for off-grid systems, you may want to check out the second edition of my book, *Power from the Sun.* Bear in mind that because batteries are expensive, it's a good idea to make your home as efficient as possible to reduce the size of your battery bank.

Battery Maintenance, Safety, and Proper Installation

Battery care and maintenance are vital to the long-term success of battery-based renewable energy systems. Proper installation and maintenance increase the service life of a battery. Longer service life results in lower operating costs over the long haul and cheaper electricity.

Keep Them Warm

Batteries like to be kept warm. In fact, batteries like to "live" in the temperature range that most human beings find comfortable. For optimal function, lead-acid batteries should be kept at around 75°to 80°F (24° to 27°C). In this range, they'll accept and deliver tons more electricity. Guaranteed! Figure 7.7 shows that, as temperatures drops, battery capacity decreases. That's because cold temperatures dramatically reduce the chemical reactions occurring in batteries. This means you'll get less electricity from colder batteries.

If you can't house your batteries in the 75°–80°F (24°–27°C) range, at least try to ensure they're housed in a room where the temperature ranges between 50° and 80°F (10° to 27°C). Rarely should batteries fall below 40°F (4.5°C) or exceed 100°F (38°C). Whatever you do, don't store

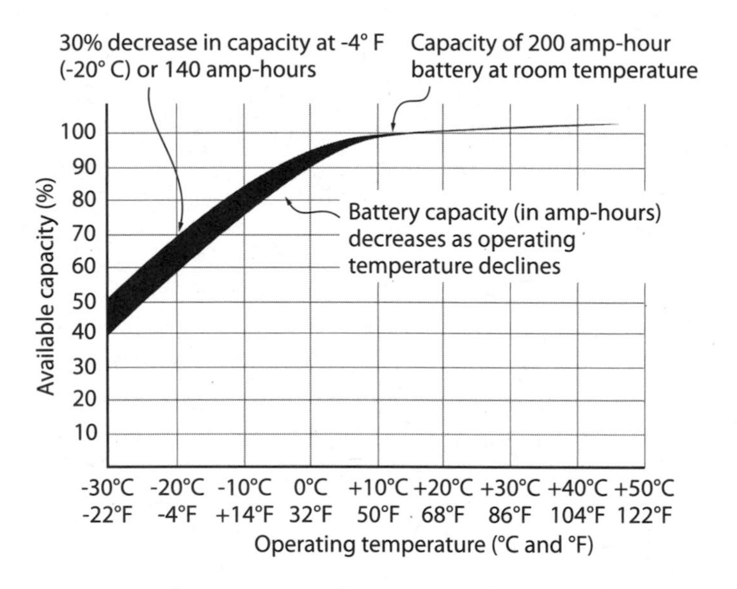

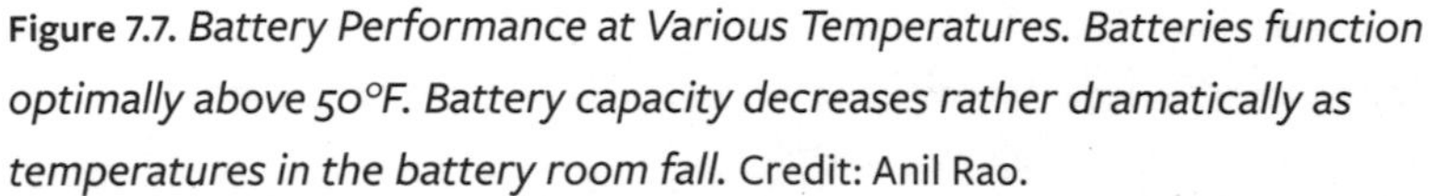
Figure 7.7. *Battery Performance at Various Temperatures. Batteries function optimally above 50°F. Battery capacity decreases rather dramatically as temperatures in the battery room fall.* Credit: Anil Rao.

batteries in a cold garage, barn, or shed. Besides delivering less electricity, they won't last long. They could even freeze under certain conditions, causing their cases to crack. This could lead to a messy and dangerous acid spill.

Batteries should not be stored on concrete floors. Cold floors cool them down and reduce their capacity. Always raise batteries off the floor. If you can, install a layer of insulation between the batteries and the floor.

Ideally, batteries should be housed in a separate, conditioned (heated and cooled) battery room or in a battery box inside a conditioned space to maintain the optimum temperature. Battery boxes are typically built from plywood. An acid-resistant liner is required to contain possible acid spills. Lids should be hinged and sloped to discourage people from storing items on top. And, as noted earlier in the book, batteries should be located as close as possible to the inverter and charge controller to minimize power losses.

Watering and Cleaning Batteries

To properly maintain batteries, you must periodically add distilled water. This replaces water lost by electrolysis, the splitting of water molecules in the electrolyte. Electrolysis occurs when electricity flows into a battery during recharge. Electricity splits water molecules in the electrolyte into hydrogen and oxygen gases. (Electrolysis is the source of the potentially explosive mixture of hydrogen and oxygen gas that makes battery room venting necessary.) These gases can escape through the vents in the battery caps on flooded lead-acid batteries, lowering water levels. A mist of sulfuric acid can also escape through the vents during charging, depleting fluid levels.

All of these sources of water loss add up over time and can run a battery dry. When the plates are exposed to air, they quickly begin to corrode. When this happens, a battery's life is over.

To prevent flooded lead-acid batteries from running dry and to ensure optimum performance, you must check battery fluid levels regularly. Many experts recommend checking batteries monthly. Others recommend checking batteries every two to three months. I recommend checking every month, just to be safe.

To check fluid levels, unscrew the battery caps and peer into each cell when the batteries are not charging. Use a flashlight, if necessary—never a flame from a cigarette lighter! Battery acid should cover the lead plates at all times—at a bare minimum a quarter of an inch above the plates. As a rule, it is best to fill batteries to just below the bottom of the fill well—the opening in the battery casing into which the battery cap is screwed.

When filling a battery, be sure to only add distilled or deionized water. Distilling and deionizing are different processes that produce similar-quality pure water. *Never use tap water.* It will contain minerals and possibly chemicals that will contaminate the battery fluid, reducing a battery's life span.

Batteries should be fairly well charged before topping them off with distilled water. So, don't fill batteries, and then charge them with a backup generator. Bear in mind, too, that overfilling a battery can result in battery acid bubbling out of the cells when the batteries are charged. If the electrolyte level is extremely low, add a little distilled water, then charge the batteries. When they are charged, you can then finish watering them.

Electrolyte loss in overly filled batteries not only reduces battery acid levels, it deposits a fine mist of acid on the surface of batteries. When it dries, the acid forms a white coating. This not only looks messy, but it can reportedly conduct electricity out of one's batteries, slowly draining them. (I'm not sure I believe this, but that's what some sources say.) Battery acid also corrodes the metal components—electrical connections, battery terminals, and battery cables.

The tops and sides of batteries need to be cleaned periodically with distilled water and paper towels or a clean rag. When cleaning batteries, be sure to wear gloves, protective eyewear, and a long-sleeved shirt—one you don't care about. If you get acid on your skin, wash it off immediately with soap and water.

When filling batteries, be sure to take off watches, rings, and all other jewelry, especially loose-fitting jewelry. Metal jewelry will conduct electricity if it contacts both terminals of a battery or a positive terminal of one battery and a negative terminal of another. Such an event will leave your jewelry in a puddle of metal—along with some of your flesh. One 6- or 12-volt cell can produce more than 8,000 amps if the positive and negative terminals of a battery are connected. In addition, sparks could ignite hydrogen and oxygen gas in the vicinity, causing an explosion. Shorting out a battery can also crack the case, releasing battery acid.

Also be careful with tools when working on batteries—for example, tightening cable connections. A metal tool that makes a connection between oppositely charged terminals on a battery may be instantaneously welded in place. The tool will become red hot and could also ignite hydrogen gas, causing an explosion. Wrap hand tools used for battery maintenance in electrical tape so that only one inch of metal is exposed on the "working" end; that way it can't make an electrical connection. Or buy insulated tools to prevent this from happening.

You may also have to clean the battery posts of your batteries every year or two. To clean the posts, use a small wire brush, perhaps in conjunction with a spray-on battery terminal cleaner, available at auto supply stores. To reduce maintenance, coat battery posts with Vaseline or a battery protector/sealer, available at auto supply stores. This protects the posts and the nuts that secure the battery cables from the acidic mist.

Ventilate Your Batteries

Batteries release potentially explosive oxygen and hydrogen gases when being charged, as just noted, so battery boxes and battery rooms should be well ventilated (Figure 7.8). Ventilating them allows these gases to escape. Never place batteries in a room near a gas-burning appliance or an

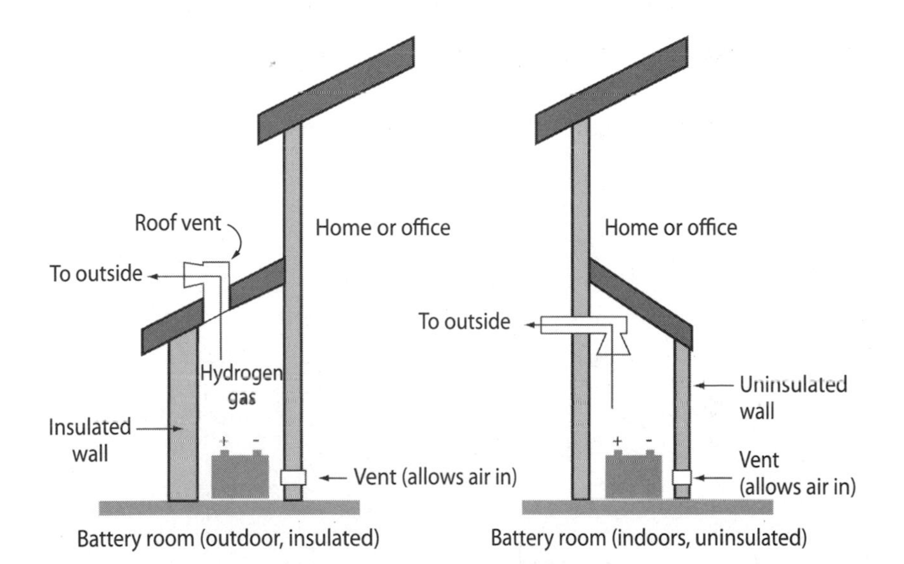

Figure 7.8. *Battery Vent System. Flooded lead-acid batteries require venting to remove hydrogen released when batteries are charging from a PV array, gen-set, or grid. Vents drawn here are supersized to emphasize their importance. In most cases, smaller passive vents are required. Be sure to allow replacement air to enter near the bottom of the battery room.* Credit: Anil Rao.

electrical device, such as an inverter, even if the enclosures are vented. A tiny spark could ignite the gases, causing an explosion.

Keep Kids Out

Battery rooms and battery boxes should be located in an out-of-the-way spot and locked, if children are present. This will prevent kids from coming in contact with the batteries and risking severe electrical shock, acid burns, and even electrocution.

Avoiding Deep Discharge to Ensure Longer Battery Life

Keeping flooded lead-acid batteries warm ensures a long life span and optimum long-term output. Longevity can also be ensured by keeping batteries as fully charged as possible. Like many technologies, lead-acid batteries last longer the less you use them. That is to say, the fewer times a battery is deeply discharged, the longer it will last. As illustrated in Figure 7.9, a lead-acid battery that's frequently discharged to 50% will last for slightly more than 600 cycles—if it is recharged after each deep discharge. If discharged no more than 25% of its rated capacity—and recharged after each deep discharge—the battery should last about 1,500 cycles. If the battery is discharged only 10% of its capacity, it will last for 3,600 cycles.

This topic (like so many others) is complicated. While deep discharging reduces the lifespan of a battery, what renewable energy users want

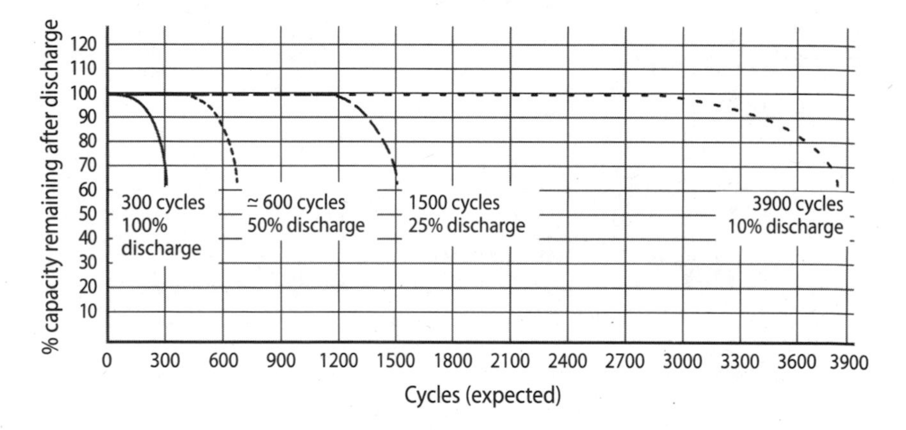

Figure 7.9. *Battery Life vs. Deep Discharge. This graph shows that shallow discharging results in more cycles. As a result, batteries will last longer if they are not discharged as deeply.* Credit: Anil Rao.

from batteries is not simply for them to last a long time but to cycle a lot of energy. We are more concerned with the cost of the battery per watt-hour cycled. So, even though it goes against the "shallow cycling makes batteries last longer rule," users have found that you'll get the most bang for your buck by cycling in the 40% to 60% deep-discharge range.

It's also very important to recharge batteries as quickly as possible after deep cycling. For long life, you should never leave batteries at a low state of charge for a long time. This results in the formation of large lead sulfate crystals on the lead plates. Unfortunately, achieving these goals is easier said than done. If your system is small and you don't pay much attention to electrical use, you'll very likely overshoot the 40% to 60% mark time and time again.

One way of reducing deep discharging is to conserve energy and use electricity as efficiently as possible. Conserving energy means not leaving lights and electronic devices on when they're not in use—all the stuff your parents told you when you were a kid. It also means ridding your home or business of phantom loads. Energy efficiency means installing energy-efficient lighting, appliances, electronics, and so on—the actions energy conservation experts have been suggesting for decades.

Conserving energy and making your home or business much more energy efficient is only half the battle, however. You may also have to adjust electrical use according to the state of charge of your batteries. In other words, cut back on electrical usage when batteries are more deeply discharged and shift your use of electricity to periods when the batteries are more fully charged. You may, for instance, run your washing machine and microwave when the Sun is shining and your batteries are full, but hold off when it's cloudy and batteries are running low—unless you want to run a backup generator.

To keep track of your batteries' state of charge so you can manage them better, it is wise to install a special meter that keeps track of how much electricity is stored in your battery bank. Some charge controllers provide this capability. If batteries are approaching the 40% to 60% discharge mark, hold off on activities that consume lots of electricity. Or, you may want to run your backup generator to recharge your batteries.

Equalization

To get the most out of batteries, you need to periodically equalize them. Equalization, mentioned earlier, is a controlled overcharge of batteries.

Why Equalize?

Periodic equalization is performed for three reasons. The first is to drive lead sulfate crystals off the lead plates, preventing the formation of larger crystals. As already noted, large crystals reduce battery capacity. Crystals can also flake off, removing sizeable chunks of lead from the plates.

Batteries must also be periodically equalized to stir the electrolyte. Because it is denser, sulfuric acid tends to sink to the bottom of the cells in flooded lead-acid batteries. During equalization, hydrogen and oxygen gases released by electrolysis create bubbles. These bubbles mix the fluid so that the concentration of acid is equal throughout each cell of each battery, ensuring better function.

Equalization also helps bring all of the cells in a battery bank to the same voltage. That's important because some cells sulfate more than others. As a result, their voltage may be lower. A single low-voltage cell in one battery reduces the voltage of the entire string. In many ways, a battery bank is like a camel train. It travels at the speed of the slowest camel.

Although equalization removes lead sulfate from plates, restoring their function, some lead dislodges—or flakes off—during equalization, settling to the bottom of the batteries. As a result, batteries lose lead over time and never regain their full capacity.

How to Equalize

Equalizing batteries is a simple process. In those systems with a gen-set for backup, the owner simply sets the inverter to the equalization mode and then cranks up the generator. The inverter controls the process from that point onward—creating a nice, controlled overcharge. In wind/PV hybrid systems, the operator can also set the controller to the equalize setting during a storm or period of high wind. The controller takes over from there.

How often batteries should be equalized depends on whom you talk to and how hard you work your batteries. Some installers recommend equalization every three months. If your batteries are frequently deep discharged, however, you may want to equalize more frequently. If batteries are rarely deep discharged, they'll need less frequent equalization. For example, batteries that are rarely discharged below 50% may only need to be equalized every six months.

Rather than second guess your batteries' needs for equalization, it is wise to check the voltage of each battery every month or two. If you notice that the voltage of one or two batteries is lower than others, equalize the battery bank. Checking voltage requires a small digital voltmeter (Figure 7.10).

Another way to test batteries is to measure the specific gravity of the battery acid using a hydrometer (Figure 7.11). Specific gravity is a measure of the density of a fluid. Density is related to the concentration of sulfuric acid in battery fluid—the higher the concentration, the higher the specific gravity. If significant differences in the specific gravity of the battery acid are detected in cells of a battery bank, it is time to equalize.

Checking the voltage of the batteries or the specific gravity of the cells may involve more work than you'd like and may not be necessary if

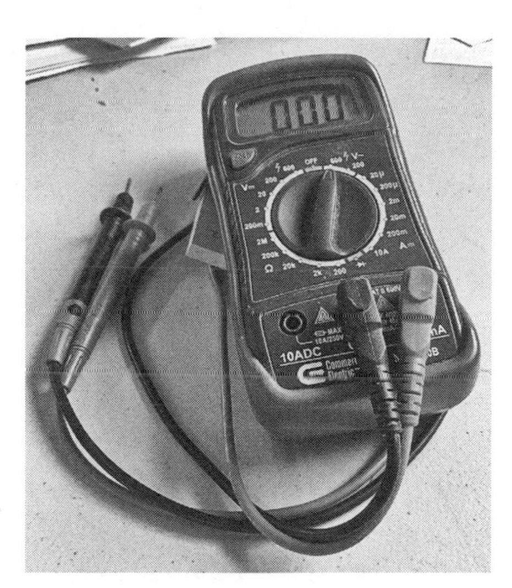

Figure 7.10. *Voltmeter for Batteries. Multimeters like these can be used to measure the DC voltage of a battery. You can purchase them at hardware stores and home improvement centers as well as electrical supply houses. Don't waste your money on an analog unit. Go digital!* Credit: Dan Chiras.

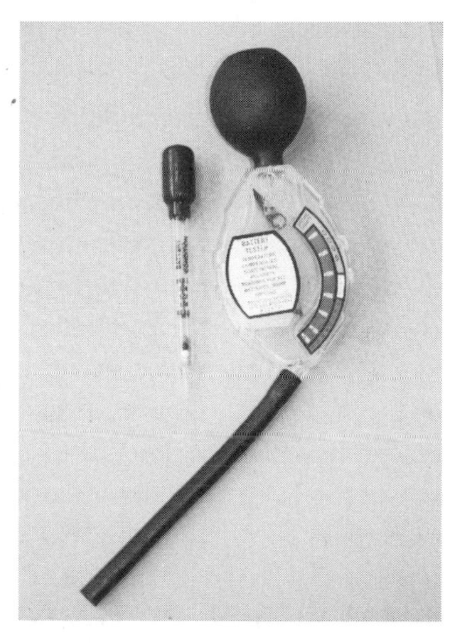

Figure 7.11. *Hydrometer. Hydrometers like these can be used to test the condition of battery acid, one cell at a time, a process that is time consuming and boring. They measure the specific gravity, which is determined by the concentration of sulfuric acid, as explained in the text. Be careful when you do this so you don't splash sulfuric acid on yourself.* Credit: Dan Chiras.

you pay attention to weather and battery voltage or adhere to a periodic equalization regime. I rarely equalized batteries during Colorado's sunny summer months, as my batteries were often full to overflowing with the electricity produced by my solar system. In the winter, however, I equalized every two months. Also, when batteries were running low, I'd often run my generator for an hour or two to bring the charge up to prevent deep cycling. This is not an equalization, just an attempt to recharge the batteries more often in cloudy weather.

As a final note on the topic, be sure only to equalize flooded lead-acid batteries. A sealed battery—either gel cell batteries or absorbed glass matt sealed batteries—cannot be equalized! If you try to, you'll ruin them.

Reducing Battery Maintenance

Battery maintenance should take no more than 30 minutes a month. (It takes about ten minutes to check the cells in a dozen batteries, but may take 20 minutes to add distilled water to each cell if battery fluid levels are low.) To reduce maintenance time, you can install sealed batteries, although they're best suited for grid-connected systems with battery backup.

Another way to reduce battery maintenance is to replace factory battery caps with *Hydrocaps,* shown in Figure 7.12. Hydrocaps capture much of the hydrogen and oxygen gases released by batteries when charging under normal operation. The gases are recombined in a small chamber in the cap filled with tiny beads coated with a platinum catalyst. Water formed in this reaction drips back into the batteries, reducing water losses by around 90 percent.

Another option is the *Water Miser cap.* These devices capture moisture and acid mist escaping from batteries' fluid, reducing water loss by around 30% to 75%.

Figure 7.12. *Hydrocaps shown here are special battery caps that convert hydrogen and oxygen released by batteries into water, which drips back into the cells of flooded lead-acid batteries, reducing watering.* Credit: Joe Schwartz.

Yet another way to reduce maintenance is to install an automatic or semiautomatic battery-filling system, shown in Figure 7.13. I've used the manually operated Qwik-Fill battery watering system manufactured by Flow-Rite Controls in Grand Rapids, Michigan, and sold online. This system worked extremely well. It turned battery maintenance from a chore to a pleasure.

Although battery-filling systems work well, they're costly. However, these systems quickly pay for themselves in reduced maintenance time and ease of operation. The convenience of quick battery watering overcomes the procrastination that leads to costly battery damage. A cheaper alternative to battery filling is a half-gallon battery filler bottle (Figure 7.14).

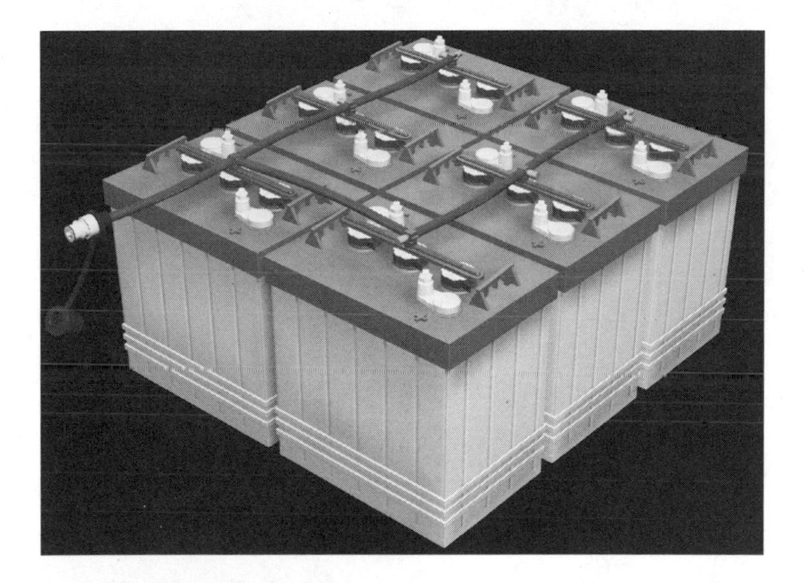

Figure 7.13. *Battery-filling System. The Pro-Fill battery-filling system greatly reduces the time required to fill batteries and makes the chore much easier.* Credit: Jan Watercraft.

Figure 7.14. *Battery Filler Bottle. Battery filler bottles work well for systems in which batteries are accessible.* Credit: Dan Chiras.

Charge Controllers

Now that you understand how batteries work and how to take care of them, let's turn our attention to the charge controller, another device that helps us care for our batteries.

A charge controller is a key component of battery-based PV systems (Figure 7.15). It performs several functions, the most important of which is preventing batteries from overcharging.

How Does a Charge Controller Prevent Overcharging?

Charge controllers monitor battery voltage at all times. When the voltage reaches a certain pre-determined level, known as the *voltage regulation* (VR) *set point,* the controller either slows down or terminates the flow of electricity (the *charging current*) into the battery bank, depending on the design. This prevents overcharging.

In some systems, the charge controller sends surplus electricity to a diversion (or dump) load (Figure 7.16). This is a load that's not critical to the function of the home or business, but is still useful. Diversion loads can be a heating element placed inside a water heater or a wall-mounted electric resistance heater that provides space heat. In PV systems, excess power is often available during the summer months during sunny periods.

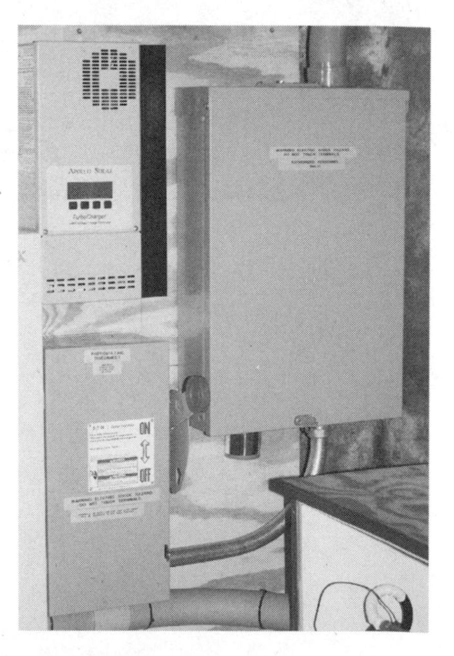

Figure 7.15. *Charge Controller. Charge controllers like the one top left from Apollo Solar regulate the flow of electricity to the batteries in off-grid and grid-connected systems with battery backup. Some charge controllers contain maximum power point tracking circuitry to optimize array output and other features as well, like digital meters that display data on volts, amps, and electricity stored in battery banks.*
Credit: Apollo Solar.

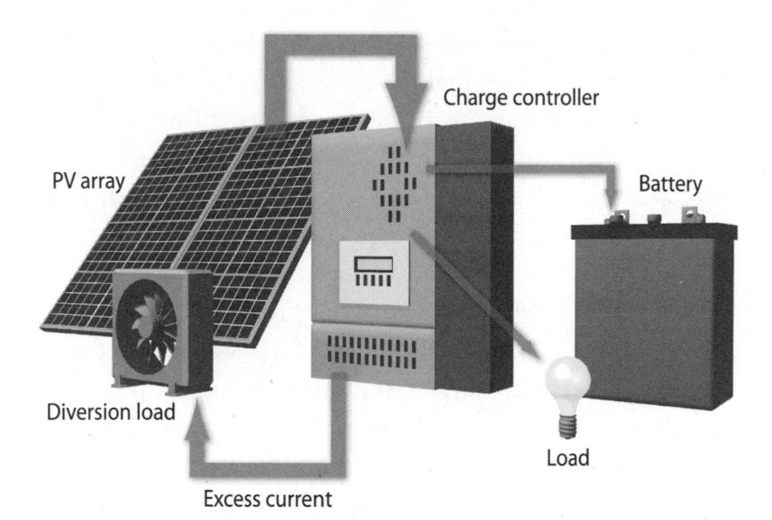

Figure 7.16. *Diversion Charge Controller. Charge controllers can be programmed to divert surplus electricity to ancillary (nonessential) loads when batteries are full, allowing a homeowner to get more out of his or her PV system.* Credit: Anil Rao.

Preventing Reverse Current Flow

At night, when the PV array is no longer producing electricity, current can flow from the batteries back through the array. Charge controllers prevent reverse current flow. When it is not prevented, reverse current flow will slowly discharge a battery bank. In most PV systems, battery discharge through the modules is fairly small and power loss is insignificant. However, reverse current flow is much more significant in larger PV systems. Fortunately, nearly all charge controllers deal with this potential problem automatically.

In these instances, the diversion load may consist of an irrigation pump or ceiling fans or fans that exhaust hot air from buildings.

Why Is Overcharge Protection So Important?

Overcharge protection is important for all lead-acid batteries in solar electric systems. That's because the current from a PV array flows into a battery in direct proportion to the amount of sunlight striking it. Problems arise when the battery reaches full charge. Without a charge controller, excessive amounts of current could flow into the battery, causing battery voltage to climb to extremely high levels. High voltage over an extended period causes severe off-gassing and water loss that can expose the lead plates to air—permanently damaging them. It can also result in internal heating and can cause the lead plates to corrode. This, in turn, will decrease the cell capacity of the battery and cause it to die prematurely.

Overdischarge Protection

Charge controllers protect batteries from high voltage, but also from overdischarging, that is, discharging too much. When the weather's cold, overdischarge protection also protects batteries from freezing. This protection is provided by an automatic *low-voltage disconnect.*

Charge controllers prevent overdischarge by disconnecting DC loads—active DC circuits in a home or business. (DC loads are common in small off-grid applications.) The low-voltage disconnect trips when a battery bank reaches a certain preset voltage or state of charge. Overdischarge not only protects batteries, it can protect loads, some of which may not function properly, or may not function at all at lower-than-normal voltages.

Generators

Another key component of off-grid systems is the generator (Figure 7.17). Generators (also referred to as "gen-sets") are used to charge batteries during periods of low insolation. They are also used to equalize batteries and to provide power for extraordinary loads—for example, welders—that exceed the output of the inverter. Finally, gen-sets may be used to provide power if an inverter or some other vital component

Figure 7.17. *Gen-set. Portable gen-sets like these commonly run on gasoline.*
Credit: Cummins Power Generation.

breaks down. Although a battery-charging gen-set may not be required in hybrid systems with good solar and wind resources, most off-grid homes and businesses have one.

Gen-sets for homes and businesses are usually rather small, around 4,000 to 7,000 watts. Generators smaller than this are generally not adequate for battery charging.

Generators can be powered by gasoline, diesel, propane, or natural gas. By far the most common gen-sets used in off-grid systems are gasoline-powered. They're relatively inexpensive and widely available. Gas-powered generators consist of a small gas engine that drives a generator. Most produce AC electricity.

Gas-powered generators operate at 3,600 rpm and, as a result, tend to wear out pretty quickly. Although the lifespan depends on the amount of use, don't expect more than five years from a heavily used gas-powered gen-set. You may find yourself making an occasional costly repair from time to time as well. To avoid these problems, size your solar electric system and battery bank properly and use energy efficiently—especially during cloudy periods, so you don't have to run your generator very much.

Because they operate at such high rpms, gas-powered gen-sets are also rather noisy; however, Honda makes some models with mufflers that are remarkably quiet. If you have neighbors, you'll very likely need to build a sound-muting shed to reduce noise levels, even if you install a quiet model. Be sure the shed is well ventilated. Don't add an additional muffler to a conventional gas-powered generator. If an engine is not designed for one, adding one could damage it.

If you're looking for a quieter, more efficient generator, you may want to consider one with a natural gas or propane engine. Large-sized units—around 10,000 watts or higher—operate at 1,800 rpm and are quieter than their less expensive gas-powered counterparts. Lower rpm also translates into longer lifespan. Natural gas and propane are also cleaner burning fuels than gasoline. Unlike gas-powered generators, natural gas and propane generators require no fuel handling by you, but they tend to be a bit more costly than comparable gas-powered generators.

Another efficient and reliable option to consider is a diesel generator. Diesel engines tend to be much more rugged than gas-powered engines. As a result, they tend to operate with fewer problems and for long periods. Diesel generators are also more efficient than gas-powered generators.

Although diesel generators offer many advantages over gas-powered generators, they cost more. And, of course, you will have to fill the tank from time to time. They're also not as clean burning as natural gas or propane gen-sets.

Conclusion: Living with Batteries and Generators

Batteries work hard for those of us who live off grid. To do their job, they require proper installation, housing, and care. As you have seen, flooded lead-acid batteries need to be kept in a warm place, but not too warm. They need to be installed in vented enclosures and must be kept clean. They also need to be periodically filled with distilled water. And, as if that's not enough, you will need to monitor their state of charge and either back off on electrical use during cloudy periods or charge them periodically with a backup power source when they're being overworked. Whatever you do, don't allow batteries to sit more than an few days in a state of deep discharge.

Generators in off-grid systems need a bit of attention, too. You will need to periodically change oil and air filters. And you'll need to fire it up from time to time to charge or equalize your batteries. It is also a good idea to run a generator from time to time during long periods of inactivity, for example, over the summer when a generator is typically not used. That's because gasoline goes bad when sitting in a gas tank for several months. (It forms a lacquer- or varnish-like substance that clogs up the works, especially carburetors.) You may also want to add a fuel stabilizer to the tank during such periods. If gasoline evaporates leaving behind varnish from the carburetor, expect a major repair bill. Finally, gasoline-powered generators can be difficult to start on cold winter days, so be sure to use the proper weight oil during the winter.

Gas-powered generators, while inexpensive, tend to require the most maintenance and have the shortest lifetime. Be prepared to haul your gas-powered gen-set in for an occasional repair.

In grid-connected systems with battery backup, you'll have much less to worry about. If you install sealed batteries, for example, you'll never need to check the fluid levels or fill batteries. Automatic controls keep the batteries fully charged.

Batteries may seem complicated and difficult to get along with, but if you understand the rules of the road, you can live peacefully with these gentle giants and get many years of faithful service. Break the rules, and you'll pay for your inattention and carelessness.

CHAPTER 8

Mounting a PV Array for Maximum Output

For a solar electric system to produce the most electricity, it must be properly installed. Break any of the rules, and you'll pay a penalty in lower production for the entire lifespan of your system. The penalty may be small, but small losses add up over the 30+ year lifespan of a solar array.

First and foremost, the array must be mounted in a location that receives as much sunlight as possible throughout the year. This can be quite challenging in urban and suburban locations. Tall trees, neighboring buildings, and chimneys can shade arrays part of the time. While every effort should be made to install a PV array in a shade-free location, compromises are sometimes necessary. As a rule, though, arrays should not be mounted on east- or west-sloping roofs unless the slope of the roof is pretty shallow—not more than 10% (Figure 8.1).

For best results, the array should be oriented so that it points directly to true south in the Northern Hemisphere or true north in the Southern Hemisphere. As noted earlier, if the tilt angle of the array is fixed, that is, it can't be adjusted, the array should be mounted at an angle equal to the latitude of the site minus 5°. For absolute best results, arrays should be installed so that the tilt angle can be manually or automatically adjusted to optimize output. (Tilt angle is most often manually adjusted, if it is adjusted at all.)

Other factors may also affect the siting of a PV array, including aesthetics, historic preservation district rules, homeowners association rules, spouses, and neighbors. If at all possible, avoid placing an array in a location that will reduce its output.

This chapter covers high-performance installation—installation that will ensure a safe, highly productive PV system. I will introduce you to

Figure 8.1. *East-facing Array. Irresponsible and uninformed solar installers sometimes install PV arrays on east- and west-facing roofs (like the one shown here) under the misguided notion that this orientation will only reduce PV output by 15%. That's not true for steep roofs like this one. This array, which faces due east, is in the shade shortly after solar noon each day. There are also several large shade trees east of the array that block sun part of the year. And, making matters worse, the east-west oriented roof of this house shades the lower part of the array on winter days when the Sun is low in the sky. Don't make this egregious mistake.* Credit: Dan Chiras.

a number of installation options that provide some flexibility in meeting the twin goals of maximum output and aesthetics.

What Are Your Options?

When it comes to mounting a solar array, you will have two options: a building-mount or a ground-mount. Building-mounted arrays are most common in densely packed areas, like cities and suburbs. In these locations, arrays typically end up on roofs so they are above trees and buildings. Modules are generally mounted on lightweight aluminum racks. Building-mounted arrays also include *building-integrated PV,* or BIPV, wherein the array is installed as roof tiles or as awnings over windows on the south side of buildings to provide shade. BIPV was discussed in Chapter 3.

Ground-mounted arrays are most common in rural areas, where there's a lot more real estate. Two options are available when mounting an array on the ground: rack-mount and pole-mount. Rack mounts are the most common.

In this chapter, I'll discuss the various options for ground and building mounting and will examine the pros and cons of each one.

Ground-Mounted Racks

My favorite way to mount PV arrays is on a rack on the ground. The main reason for this is that ground-mounted arrays tend to stay cooler in the summer than roof-mounted arrays. Cooler arrays produce more electricity because high heat decreases the output of PV modules.

In a ground-mounted PV array, modules are attached directly to extruded aluminum rails like those featured in Figure 8.2. As illustrated, the aluminum rails have channels that run their full length. These channels allow installers to attach modules and also to attach the rails to the rest of the rack. One innovative rail that I use to cut down on installation time on ground-mounted PV systems is manufactured by PV Racking in Pennsylvania. As shown in Figure 8.2b, it consists of two large channels

Figure 8.2a, b, and c. *Several Types of Rails. Rails for mounting PV modules consist of a series of channels used to attach them to modules and mounting hardware, typically L-feet. (left) Most rails on the market look like this one from Renusol, and most modules are attached by midclamps and endclamps, (not shown). PV Racking makes a rail that I use a lot. As you can see, it has slots into which installers slide the modules. No WEEBs or ground lugs and ground wire are required. (right) A single worker can install the modules by himself.*
Credit: Dan Chiras.

Figure 8.3. *Note the cement blocks (ballast) that secure the array on the roof.* Credit: Dan Chiras.

into which an installer slides the modules. This feature greatly reduces installation time and cost (Figure 8.2c). Rails are attached to an aluminum or steel substructure (Figure 8.3). Together, the rails and the substructure form the rack.

To remain upright in strong winds, racks must be secured to the ground. That's usually accomplished by attaching them to a steel-reinforced concrete foundation, like the one shown in Figure 8.3. This adds cost to an installation, but the benefits of ground mounting often outweigh the additional cost. In parking lots, PV arrays can be anchored to large aboveground concrete blocks. (You'll see a photo of this shortly.)

Arrays can also be secured to the ground by attaching them to galvanized steel augers, as shown in Figures 8.4 and 8.5. These are screwed into the ground and the steel rack is then attached to them.

Ground-mounted racks come in two basic varieties: fixed and adjustable. Fixed racks are typically set at the optimum angle (the latitude of a location minus 5°). Adjustable racks can be adjusted to maximize annual energy production by an array. Most arrays are adjusted twice a year. Adjustability is typically provided by telescoping back legs. Changing the length of the back legs allows the tilt angle to be increased or decreased (Figure 8.6). Some telescoping legs come with predrilled holes or slots that correspond to different tilt angles. This limits your options. Most legs I've used on ground and roof

Figure 8.4. *Steel Augers. Galvanized steel augers like the one shown here can be used to anchor a PV array to the ground.* Credit: Dan Chiras.

Figure 8.5. *Steel Augers Being Installed. We use this machine or a skid steer equipped with an auger attachment to install augers to mount PV systems. One thing I like about this system is that it minimizes ground disturbance and makes for a much neater installation.* Credit: Dan Chiras.

Figure 8.6. *Telescoping Legs. Telescoping legs, like the ones on TEI's first solar array in 2009, allow for infinite adjustability within a range dictated by the length of the two aluminum pieces, as shown here. Here, students and a veteran installer (Bob Solger) who helped teach this workshop check the angle and distance from the front to the back rail.* Credit: Dan Chiras.

mounts, however, are infinitely adjustable. A telescoping leg consists of two pieces of aluminum. The smaller piece (C-channel) that slides inside the larger-piece (square tubular aluminum) as illustrated in Figure 8.6. A single bolt secures the legs at any point you want.

Figure 8.7. *Pole-Mounted Array. Each array is mounted on a pole anchored to the ground and is installed at the optimum angle for its location. Pole-mounted arrays provide some flexibility in placing an array on a lot.* Credit: Anthony Powell.

Ground-Mounted Pole-Mounts: Fixed and Tracking Arrays

Figure 8.7 shows two pole-mounted arrays. As shown in Figures 8.8 and 8.9, the modules are attached to an aluminum rack that's mounted on a sturdy steel pole. The poles are typically embedded in a steel-reinforced concrete base. PV arrays are usually

Figure 8.8. *Pole-Mount Being Assembled. Students at MREA workshop installing rails on the strong back (center beam) on a pole-mount.* Credit: Dan Chiras.

Figure 8.9. *Module Being Inserted in Pole-Mount. Workers carefully slide PV modules into place on this dual-axis tracker at a Midwest Renewable Energy Association workshop in Minnesota.* Credit: Dan Chiras.

mounted on the top of the pole. Top-of-pole-mounts allow operators to seasonally adjust the tilt angle of the array to accommodate changes in the angle of the Sun.

Pole-mounted arrays fall into two categories: fixed or adjustable. Fixed arrays are immovable. Modules are mounted so that they point to true south at a fixed angle.

There are two types of adjustable pole-mounted arrays: manually or automatically adjustable. A manually adjusted array allows the owner to change the tilt angle of the array as often as he or she desires. Some homeowners adjust their arrays four times a year. Others adjust twice—in the spring and fall. In the spring, the tilt angle of the array is decreased by 15° below the optimum angle to capture more high-angle summer sun, and in the winter, it is raised by 15° to capture more low-angle winter sun.

Seasonal adjustments of the tilt angle can increase the output of an array by 10% to nearly 40%, depending on the latitude of the site, location of the array on one's property, and shading.

Adjusting a very small, ground-mounted array may be as simple as loosening a nut on a bolt on the back of the array mount, then tilting the array up or down (Figure 8.10). A magnetic angle finder, shown in Figure 8.11,

Figure 8.10. *Adjustment for Pole-Mounted Array. The tilt angle of a seasonally adjustable pole-mounted array can be changed to accommodate the changing altitude angle of the Sun to increase electrical production.* Credit: Dan Chiras.

can be used to set the tilt angle. Once set, the nut is tightened. Manually adjusting larger and more commonly used pole-mounted arrays, however, can be a challenge. The task often requires three or four people.

Another option for a pole-mounted array is a *tracking array.* Tracking arrays are designed to track the Sun from sunrise to sunset (Figure 8.12). Two types of trackers are available: single-axis and dual-axis.

A single-axis tracker adjusts for the position of the Sun in the sky relative to true south, which is referred to as the *azimuth angle.* Single-axis trackers

Figure 8.11. *Magnetic Angle Finder. This inexpensive device is used to check the angle of roofs and arrays.* Credit: Dan Chiras.

Figure 8.12. *Tracking Array. This tracking array was installed by Dan and fellow students at a workshop in Minnesota sponsored by the Midwest Renewable Energy Association located in Custer, Wisconsin.* Credit: Dan Chiras.

therefore rotate the array from east to west, following the Sun from sunrise to sunset. The tilt angle of the array, however, remains fixed, although it can be manually adjusted two to four times a year to account for seasonal differences in the altitude angle of the Sun (the position of the Sun in relation to the horizon).

A tracker that moves the array so that it adjusts for changes in both the altitude angle and the azimuth angle of the Sun is known as a *dual-axis tracker.* Figure 2.9 shows the position of the Sun during the year, illustrating change in the altitude and azimuth angles.

Dual-axis trackers are most useful at higher latitudes because days (hours of sunlight) are much longer in the summer in such regions. In the tropics, where daylength remains the same throughout the year, a single-axis tracker will perform as well as a dual-axis tracker, provided it has access to the Sun from sunrise to sunset.

Although trackers increase an array's annual output, the greatest impact on energy production occurs during the summer, when days are longer and sunnier. Output improves less during the short, often-cloudy days of winter.

For those installing grid-tied systems, a tracking array's excess summer production helps offset winter's lower production—but only if you have annual net metering or monthly net metering that reimburses you at retail. A tracking array on a system where the utilities practice monthly net metering that reimburses at wholesale (avoided) cost won't help a whole lot, because summer surpluses (net excess generation) are typically reimbursed at one-fourth of the electricity's value.

Trackers are also less useful for off-grid systems because summer surpluses are usually wasted. Once the batteries are full, the array's output has nowhere to go. As you will recall from Chapter 7, charge controllers prevent the electricity from overcharging the batteries or, in the case of diversion charge controllers, they dump the surplus into nonessential loads.

Trackers operate either actively or passively. Active systems rely on electric motors to adjust the array's angles. In many tracking arrays, the Sun's position is detected by photosensors mounted on the array (Figure 8.13). These sensors send signals to a small computer that controls the motors (Figure 8.14). In some systems, tracking is controlled by a computer that's programmed with the altitude and azimuth angles of the Sun for every day of the year.

Figure 8.13. *Photo Sensor on Pole-Mounted PV Tracker. This electronic eye sends signals to the controller which adjusts the tilt angle and azimuth angle of the array to track the Sun across the sky.* Credit: Dan Chiras.

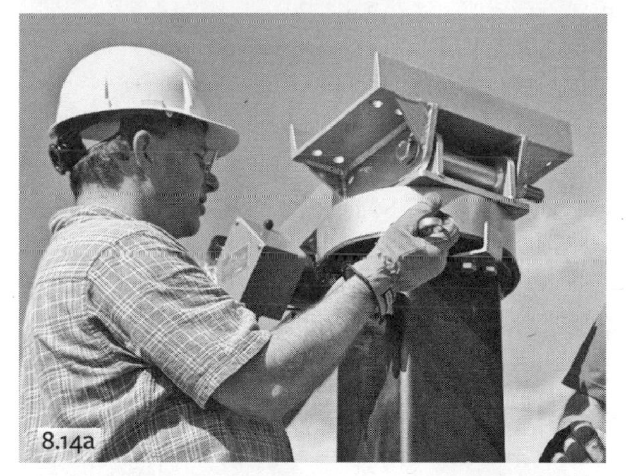

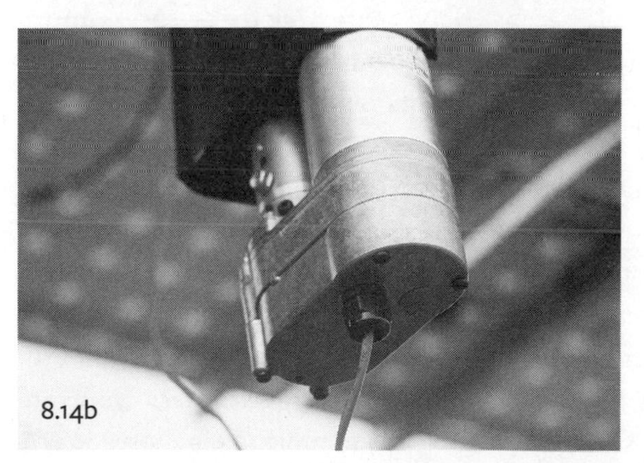

Figure 8.14. (a) *Motor That Adjusts Azimuth Angle. This motor sits atop the pole and adjusts the azimuth angle of the array to track the Sun on its east-to-west path across the sky.* (b) *Motor That Adjusts Tilt Angle. A smaller electric motor adjusts the tilt angle of the array to accommodate the ever-changing altitude angle of the Sun.* Credit: Dan Chiras.

Figure 8.15. *Passive Tracker. Passive trackers like this one do not require motors or photo sensors. How they work is explained in the text.* Credit: Zomeworks.

Passive-tracking systems are single-axis trackers that require no sensors or motors. As shown in Figure 8.15, passive trackers are equipped with fluid-filled tubes positioned on either side of the array. They are connected by a small-diameter pipe. When sunlight strikes the tube on the left, for example, right after sunrise, it heats the liquid, causing it to vaporize (expand). Expansion, in turn, forces the gas to flow into the tube on the right. This causes the weight to shift to the right, which causes the array to shift, tracking the Sun as it "moves" across the sky.

Both active and passive trackers increase array output, but active trackers are more efficient. That's because they rotate back to the east (sunrise position) at the end of each day. Passive trackers follow the Sun from sunrise to sunset but, at the end of the day, remain pointing west. It's not until sunrise that they slowly move back into position. During this time, the array produces little, if any, electricity. It can't. The array has its back to the Sun. Because a passive-tracking array must return to its east-facing position at sunrise, it begins producing electricity an hour or two later each day than an active-tracking array.

For optimum performance, arrays equipped with trackers must receive dawn-to-dusk sun. "There's no point in buying a tracker if your site doesn't begin receiving sunlight until 10 in the morning, or loses it at 2 in the afternoon due to shading from hillsides, trees, or buildings," notes Ryan Mayfield in an article in *Home Power Magazine.* So, which should you buy—a single—or a dual-axis tracker?

Even though a dual-axis tracker may seem as if it would produce a lot more electricity than a single-axis tracker, the benefit is actually only marginal. As shown in Figure 8.16, a single-axis tracker generally results in the greatest improvements in array output. Dual-axis trackers increase output over a single-axis, but only very slightly.

Should you buy an active or a passive system? The answer to this question depends on where you live. Active trackers work well in both cold and hot climates and everything in between. Passive trackers, on the other hand, do not perform as well in cold climates. They tend to be sluggish and imprecise on cold winter days. That's because they depend on solar heat to vaporize the refrigerant. If you live in a cold climate, consider an active tracker. If you live in a warmer climates, passive trackers will perform better.

Although active trackers have many advantages, they do require photosensors, controllers, and electric motors. The more parts, especially electronic parts, the more likely something will go wrong. The weakest link in active-tracking arrays is the controller. The controller is a small logic board that integrates information from the photosensors. Unfortunately, this circuitry is vulnerable to nearby lightning strikes, because they produce electromagnetic pulses that can fry electronic circuitry. While manufacturers have improved lightning protection in active trackers, a nearby or direct lightning strike can damage a controller, necessitating replacement.

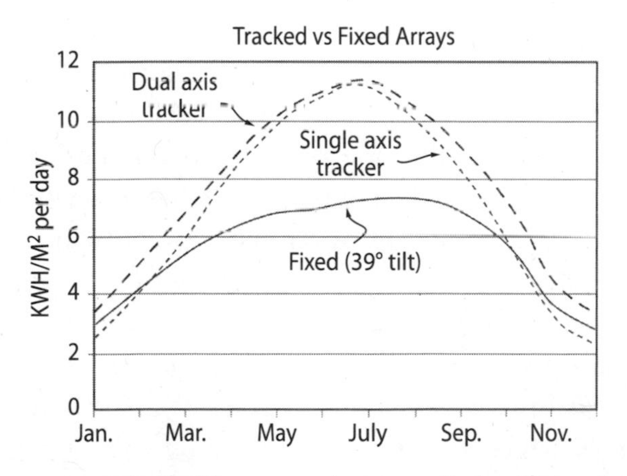

Figure 8.16. *NREL Graph Showing Output of Tracking Arrays. This graph compares the monthly output of a fixed array, a single-axis, and a dual-axis array. A single-axis tracker dramatically increases the annual output compared to a fixed array. A dual-axis array results in very little additional gain.* Credit: Anil Rao.

According to some "experts," the best answer to the question, "Should I install an active or passive system?" may be: neither. Although it is true that trackers increase the output of a PV array, the key question to ask is how much does this additional output cost? And, is it worth it? Or, are there other, perhaps more reliable and cost effective, ways to boost the electrical output of a PV array?

Proponents of the view that "no tracker is the best tracker" point out that, in most cases, it is more economical to invest in a few more modules to boost the output of a PV array than to install a tracker array that yields the same output. A larger, fixed array will require less maintenance, too, lowering its overall cost. If you're thinking about installing a tracker, run the math, or ask your installer to run it for you, to determine whether it makes sense. Be sure to take into account periodic controller replacement.

Installing a Pole-Mounted Array

To install a pole-mounted array, you will need to excavate a deep hole for its foundation. Exact specifications for the excavation and size of the pole should be provided by a pole-mount's manufacturer.

To give you an idea what's involved: for a 120-square-foot array (about 3,000 watts of solar), one pole-mount manufacturer (Direct Power and Water) recommends using 6-inch schedule 40 pipe. (Schedule 40 refers to the thickness of the pipe.) For a pole that's 6 to 7 feet (1.8 to 2.1 meters) above ground, they recommend embedding 4 to 6 feet (1.2 to 1.8 meters) of pipe in a steel-reinforced concrete foundation.

If installing a system yourself, be sure to check with the local building department. Some require a licensed engineer to specify foundation details based on local soil and wind conditions. If you're hiring a professional installer, they will handle these details when they secure the permit. If you use round pipe, be sure to secure the pipe to the reinforcing steel in the foundation so it will not twist when the winds blow. For more details regarding the installation of pole-mounted arrays, you may want to pick up a copy of my book, *Power from the Sun*.

Once the foundation has completely cured (don't rush it!), the array is mounted on a rack attached to the pole. The rails of racks are designed for specific modules, so when ordering a pole-mount rack and hardware be sure to specify the module details and the number of modules you'll be using. Some installers recommend purchasing a larger rack than initially required so more modules can be added at a later date, should you decide to increase the size of your system.

Pros and Cons of Ground-Mounted Arrays

Ground-mounted arrays may be your only choice if a roof is unsuitable for some reason—for example, if the roof faces the wrong direction, is at an inappropriate pitch, is shaded, or is not strong enough to withstand the weight of the array (that's very rare). Even if roof-mounting is an option, you'll find that ground-mounted arrays offer many advantages.

One of the biggest advantages of ground-mounted arrays is they can be positioned far away from anything that might cast a shadow on them. I typically install most grid-tied arrays within 100 to 300 feet of the main service panel in a sunny field or lawn next to homes. That said, I have installed grid-tied PV arrays as far as 1,000 feet (300 meters) from homes. (We had to use very large wire [4/0] to minimize voltage drop.)

Ground-mounted arrays can also be positioned more precisely than many roof-mounted systems. That is, they can be oriented to true south and tilted at just the right angle to ensure maximum production. Ground-mounted arrays also help to maintain cooler module temperatures than roof-mounted arrays. Most PV modules produce more electricity at cooler temperatures. Precise positioning and cooler array temperatures increase the output of a PV array and the value of your investment in PVs. In other words, you'll get more electricity out of the system over its lifetime. This, in turn, lowers the cost per kilowatt-hour. Ground-mounted arrays are also often a lot safer to install than roof arrays. There's no need for potentially dangerous roof work. Another advantage is that ground-mounted arrays do not require roof penetrations. Driving lag screws into a roof to install a rack and cutting holes in a roof to run wires to the balance of the system create potential leaks.

Ground mounting an array, as opposed to a roof mount, also avoids the dismantling of an array when time comes to reroof a house—for example, after a hail storm. Generally speaking, PV modules will outlive most roofs, sometimes multiple reroofings.

Ground-mounted arrays also generally permit easier access to solar modules than roof-mounted arrays. This enables an owner to clean a dusty array, if necessary, or gently brush snow off modules to maximize output. (Though cleaning is rarely required in most regions.) Ground mounting also permits easy access for inspection and maintenance, although this is rarely necessary. PV arrays require very little, if any, maintenance over their long lifespan.

On the downside, ground-mounted arrays are not usually suitable for small lots in cities and suburbs. Nearby homes and trees often block the Sun in backyards.

Ground-mounted arrays are also more accessible to vandals than building-mounted systems. In addition, precautions need to be taken to prevent livestock, such as horses and cattle, from using your $30,000 PV array as a scratching post. (This is usually solved by enclosing an array with an electric fence.) In more developed areas, local building departments may require fencing to keep kids out (Figure 8.17).

Building-Mounted PV Arrays

For many years, PV arrays were mounted on racks with back legs that elevated the rack from the roof, as shown in Figure 8.18. These racks, referred to as *elevated roof racks,* allowed installers to set the correct tilt angle. As solar became more popular, however, more and more building-mounted arrays were installed on standoff mounts like the one shown in Figure 8.19. Standoff mounts minimize the visual impact of PV arrays, making them

Figure 8.17. *Array Enclosed in Fence. This array at the First Methodist Church in Sedalia, Missouri, is protected by a chain link fence to keep kids and vandals out. Local building departments may require similar protection in urban and suburban settings. Notice the use of precast concrete blocks to anchor the array to the ground and ensure that it won't be blown away by the wind.* Credit: Dan Chiras.

more aesthetically appealing to homeowners, neighbors, and city officials and less vulnerable to wind.

Figure 8.18. *PV Array on Standard Rack. In early days of PV industry, modules were mounted on racks like the one shown here. They are designed to allow installers to mount modules at the proper tilt angle to maximize output, though they may not be the most attractive way to install a PV array.* Credit: Dan Chiras.

Figure 8.19. *Standoff Mount. Standoff mounts allow modules to be mounted parallel to the roof, permitting PV systems to blend in better, a feature that is desirable to homeowners, business owners, neighbors, and passersby. This array was installed by me, my business partner, and a former student on Victor Pipe and Steel in Winfield, Missouri.* Credit: Dan Chiras.

Standoff racks allow installers to mount PV arrays parallel to the roof surface to reduce their profile and visibility. Because the array is mounted parallel to the roof surface, the modules are mounted at a fixed tilt angle corresponding to the angle of the roof.

In standoff mounts, the array is typically raised off the roof by about six inches (15 cm). The space between the array and the roof allows for some air movement, but nowhere near as much as an elevated roof rack.

Installing a Building-Mounted PV Array

Elevated roof racks require the same hardware discussed earlier to install ground-mounted racks: rails to attach the modules, and back legs to angle the modules appropriately. They also require waterproof attachments to secure them to roofs. Figure 8.20 shows the most commonly used attachment, an aluminum L-foot. To prevent leakage, a high-grade roofer's caulk is applied. Installers can also use metal flashing at these roof-anchor sites to prevent leakage (Figure 8.21).

Standoff mounts require rails and six-inch aluminum posts (standoff mounts) to raise the racks off roofs (Figure 8.22). Rails are attached to the standoffs by L-feet.

Installing both types of racks on roofs requires special skills and knowledge of roof design and construction. Workshop experience and the help of experienced friends can make the job a lot easier—and go a lot faster. For best results, consider hiring a qualified installer—one who has

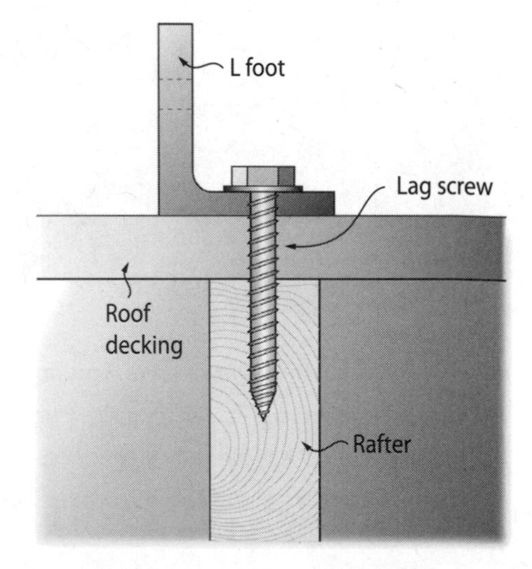

Figure 8.20. *Lag Screw. L-feet or similar attachments are lagged into rafters or top cords of roof trusses with lag screws. Manufacturers specify how far apart the L-feet should be and the size and length of the lag screws, so be sure to check. Remember, too, you are only lagging into a 1.5-inch-wide framing member, so chances of missing it or running the lag at the edge are pretty high. Always drill a pilot hole first.* Credit: Anil Rao.

installed a lot of systems. They can do the job in a fraction of the time, and the result will probably be much better.

Modules are secured to the rack in one of two ways: either from the front (top-down), or from the back. Top-down mounting requires

Figure 8.21. *Mounting Kit with Flashing. We used this roof mount with flashing on a job on a steel company in Winfield, MO. Flashings such as this help reduce the chances of leakage, giving the homeowner and the installer a little more peace of mind.* Credit: Dan Chiras.

Figure 8.22. *Standoff Mount. Standoff mounts like the one shown here allow modules to blend with the architecture, making them more aesthetically appealing to many individuals.* Credit: Dan Chiras.

mounting clips like those shown in Figure 8.23. They are connected to slots in the rails via stainless steel bolts. In backside mounts, bolts run through the rail into holes in the frame of the module. This approach is rather difficult and slow, so, for ease of installation, most installers use top-down mounting clips, called *midclamps* and *end clamps.*

When installing a PV array, be sure that the array and the racks—and the rest of the system—are properly grounded. Grounding is vital for safety and system performance. It protects users from shocks and helps protect equipment from damage. Shock may occur if insulation on a wire carrying current is damaged because it can allow the internal conductor (aluminum or copper) to come in contact with metal, for example, the frame of a module. If this occurs, the module frame will conduct electricity. If a person were to touch the metal, he or she could receive a shock.

Figure 8.23. (a) *Midclamps Attached to Rail With WEEB. Midclamps, like the one shown here, connect the modules to the underlying rail.* (b) *A WEEB is used to bond the module frame to the rail.* Credit: Dan Chiras.

Ground wires are typically 4- or 6-gauge bare copper wire. They connect the metal components like the module frames and rails to a ground-rod driven 8 feet (2.2 meters) into the ground (Figure 8.24). Copper is an excellent conductor of electricity. People are great conductors of electricity, too. We are filled with electrolyte in the form of watery fluids containing positively and negatively charged ions. Fortunately, electricity prefers the path of least resistance. And, even more fortunately, copper ground wires provide a lower-resistance pathway for electrical current to travel to the ground than your body. If the system is not properly grounded, however, your body could provide an excellent path to ground. For more details on

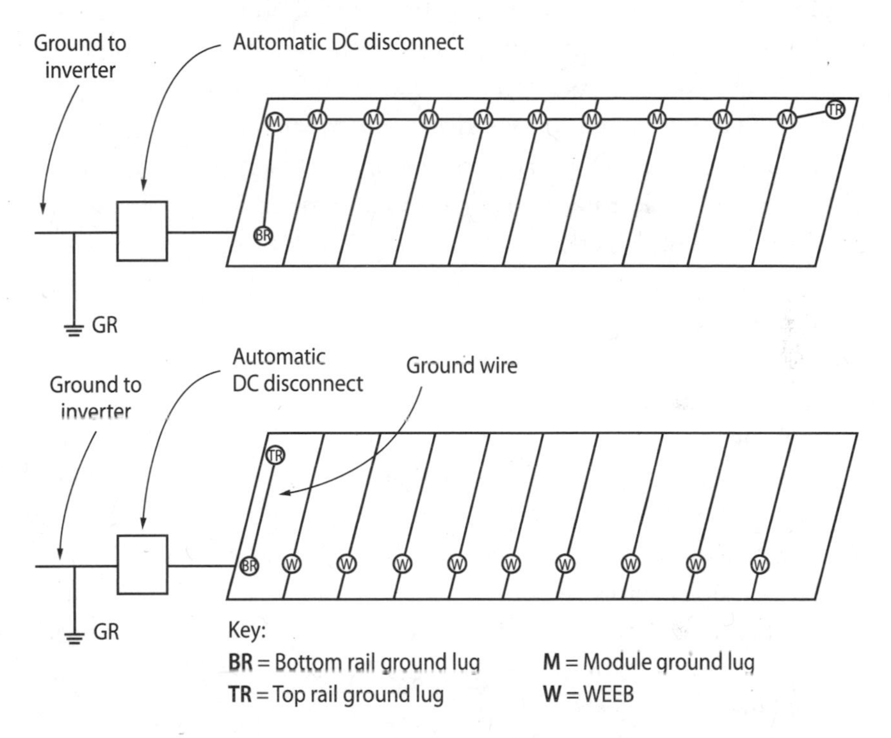

Figure 8.24. *Grounded Array. Grounding a PV system is very important and most often not done correctly.* (a) *In this drawing, a ground wire is run from module to module but also connects to the top and bottom rails of each rack, forming a continuous ground. The ground wire terminates in a combiner box at a bus bar. A new ground wire runs from the combiner box to a ground-rod or to the inverter and main service panel, where the DC and AC systems can share a common system ground. Both are perfectly acceptable according to Code.* (b) *Shows how WEEBs can replace long ground wire runs.* Credit: Forrest Chiras.

grounding, you may want to crack open a copy of my book, *Power from the Sun.*

Pros and Cons of Building-Mounted Arrays

One advantage of roof-mounted racks is that they are, in most cases, slightly easier and less costly to install than ground-mounted racks. Another advantage is that they allow installers to position PV arrays well above obstructions, such as trees or neighbors' homes. This helps boost an array's output.

Traditional elevated roof racks permit air to circulate around arrays, keeping them cooler and increasing their output. This type of rack also gives ready access the back of the modules, which makes wiring and troubleshooting easier, such as when you need to test the output of individual modules. Standoff mounts provide less room for air circulation and won't operate as efficiently. They're also much more challenging to troubleshoot.

On the downside, roof-mounted racks result in multiple roof penetrations. If not sealed properly, these can leak. Water leaking through the roof can dampen ceiling insulation, dramatically reducing its R-value. (R-value is a measure of insulation's ability to resist heat flow.) Moisture also promotes mold growth and can discolor and damage ceilings. If the leak is persistent, it can cause roof decking and framing members to rot.

Roof-mounted PV arrays are also more difficult to access than ground-mounted or pole-mounted PV arrays. This makes it more challenging to adjust the tilt angle, brush off snow, or wash off dirt and bird droppings. That said, most roof-mounted racks are set at a fixed tilt angle, and dust is rarely a problem.

Another downside of roof-mounting a PV array is that it can limit an array's exposure to the Sun if the roof of a home doesn't face true south. This will lower output.

One of the most significant disadvantages of roof-mounted racks is that they must be disassembled and removed when time comes to reroof a house. This is costly and time-consuming work that should be carried out by a trained professional—not your local roofing contractor. If your roof needs replacement, or will need replacement soon, do it *before* installing a PV system.

When building a new home or reroofing an existing home, I recommend installing a long-lasting roofing material. Metal roofs should last for

50 years or more. Environmentally friendly roofing materials made from recycled plastic and rubber may also outlast standard asphalt shingle roofs by decades. At the very least, chose high-quality (long-lasting, architectural) asphalt shingles. They're more expensive, but they are designed to last about 40 years, depending on conditions, outlasting standard asphalt roof shingle (which usually last about 20 years).

Ballast Racks

Homeowners or business owners whose roofs are "flat" (in actuality, they are not flat; they slope enough to allow water to run off) usually prefer not to penetrate their roofs when having PV systems installed. In such cases, modules can be installed on lightweight racks that rest on top of roofs. They are typically weighted down by cement blocks (the ballast) (Figure 8.25).

Ballast racks are made from lightweight aluminum or, in some cases, UV-resistant plastic. Most racks are equipped with ballast trays (or some similar structure) into which the cement blocks are placed. The trays raise the ballast off the roof to prevent damage. Modules are secured to the racks. Racks are typically installed in multiple rows, as shown in Figure 8.25.

Ballast racks offer several important advantages over other roof racks. First, they generally come preassembled, so installation is a snap. As noted earlier, they do not require roof penetration. Lack of penetrations therefore greatly reduces the chances of leakage.

Figure 8.25. *Integrated PV Ballast Mount. Note how the racks from one row are connected to the rack in the next.* Credit: Dan Chiras.

There are some downsides to ballast mounts, however. One is that a great amount of ballast must be transported to the roof and carried to the ballast trays. Roofs must be strong enough to support the weight of the ballast (though this is usually not a problem).

Another downside is that ballast racks are typically designed with a rather low 10° tilt angle to prevent them from being lifted off roofs by wind. (Steeper angles would be possible, but you'd need more ballast.) Their low tilt angle reduces array output in the winter, when the sun is low in the sky. However, it increases array output in the summer. Over a full year, it slightly decreases the electrical output of a solar array.

Railless Rack Systems

To simplify and reduce the cost of installation, rack manufacturers have developed railless rack systems. The first of these was the S-5! clamp system for standing seam metal roofs, a special kind of metal roof (Figure 8.26). Several manufacturers have created similar mounts for shingled roofs and metal roofs. These mounts are screwed into the rafters. They then attach directly to the frames of the modules, reducing cost and installation time.

Railless mounting systems reduce the amount of aluminum required for a solar array. Aluminum is a high-embodied energy material. Using

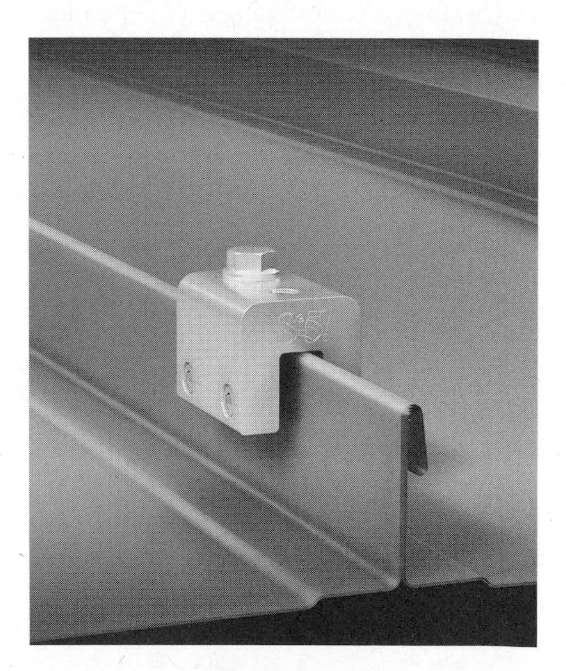

Figure 8.26. *Standoff-Mounted PV Array S-5! Clamps. The S-5! clamp shown here allows an installer to attach a PV array to a standing seam metal roof. Rails can be attached to the clamps, then modules can be attached to the rails. Railless mounts are also possible with this mounting hardware. In such cases, the modules are mounted directly to the S-5! clamps, and no rails are used.* Credit: S-5!

a railless system, therefore, not only simplifies installation, and reduces costs, it results in less energy consumption and less air pollution.

Building-Integrated PVs

In Chapter 3, I discussed the options for integrating PVs into buildings—for example, in roofs, walls, and glass. For most homeowners and businesses, there are three practical options: solar roof tiles, PV laminates on standing seam metal roofs, and solarscapes. Let's take a brief look at each one.

PV Laminates For Standing Seam Metal Roofs

If your home or business is fitted with a standing seam metal roof (SSM), or if you're about to reroof and install a standing seam metal roof, you can install a product known as *PV laminate* (PVL) (Figure 8.27). PV laminate is fabricated by applying multiple layers of amorphous silicon to a durable, but flexible metal backing. The amorphous silicon is then coated with a UV-resistant plastic. PVL is installed between the standing seams of SSM roofs, and can be applied to new or existing roofs, though, it's best applied when installing a new roof.

Installing PV Laminate

PV laminate (PVL) comes in rolls with a sticky back. They're installed one at a time on sections of metal roofing like contact paper. (It's best to install this product when the SSM roof material is on the ground, not the roof.)

Figure 8.27. *PV Laminate from Uni-Solar. PV laminate is applied directly to standing seam metal roof as shown here.* Credit: National Renewable Energy Laboratory.

When applying PVL, care should be taken to keep the laminate as straight as possible as it is unrolled. The adhesive is pretty strong. You can't lift and reposition PVL very easily if you mess up. Be sure to avoid creasing.

After the PVL is attached, the metal roof panels can be installed. The positive and negative leads from the PV laminate are then wired in series to boost voltage. Leads are then run under the ridge cap or under eaves.

Pros and Cons of PVL

In skilled hands, PVL is fairly easy to install. It blends with the building and, from a distance, may not be visible at all. Unlike conventional crystalline PV, PVL is made from amorphous silicon, which is considerably less vulnerable to high temperatures. As a result, PVL's output doesn't decrease as much at high temperatures as monocrystalline and polycrystalline PVs. PVL is also not as sensitive to shading as standard PV modules. On the downside, because PVLs that are less efficient than crystalline PVs, you may need twice as much roof space to generate the same amount of electricity as you'd get from standard crystalline modules. Standing seam metal roofing is also not as attractive, at least on homes, as other metal roofs. In fact, it is rarely used on homes.

Solar Roof Tiles

Individuals interested in BIPV, may also want to consider solar roof tiles. Solar roof tiles incorporate crystalline silicon cells in a tile that's attached directly to the roof (Figure 8.28).

Solar roof tiles do not need to cover the entire roof. You can, for instance, dedicate a portion of a south-facing roof to them. The size of the system depends on the solar resource, roof pitch, shading, and electrical demand. As in all solar electric systems, the ideal location is a south-facing roof; however, solar tiles can also be mounted on roofs that face southeast or southwest. The decline in production will vary depending on how far off from true south the roof is oriented. Mounted on east- and west-facing roofs, the output could easily decline by 10% to 15% for shallower roofs. For really steep roofs, the decrease in performance could be as high as 40% or 50%.

Solar roof tiles overlap like shingles to shed water from the roof. Most products cannot be mounted on flat roofs. Be sure to check the pitch of your roof and manufacturers' recommendations before committing to solar roof tiles.

Figure 8.28. *Sunslates from Atlantis Energy Systems. Sunslates from Atlantis Energy Systems are solar tiles that replace ordinary roof shingles. The top of this roof is fitted with solar tiles that generate electricity to help meet household demands.* Credit: Atlantis Energy Systems.

Solar roof tiles work well in a variety of climates, but perform best in sunnier locations. Be sure to check with the manufacturer for advice on the suitability of their product for your location. Also, be sure to check the compatibility of a solar tile with your roof. Sunslates, for instance, are "fully compatible with any other roof, be it tile, shake, metal or composite," according to the manufacturer. The manufacturer may also have recommendations for the best combinations. Atlantis Energy Systems, for instance, states that their product goes best with gray concrete roof tiles.

Installing Solar Roof Tiles

Solar roof tiles are mounted on roofs over 30-pound roofing felt or similar roofing materials made from plastic. Each manufacturer provides directions. As a rule, though, solar roof tiles are mounted on metal or wooden structures attached to the roof decking. They are then wired together to form series strings.

Pros and Cons of Solar Roof Tiles

Solar tiles offer many of the advantages of solar roofing products. The main benefit is aesthetics. They also have some of the same downsides, the most important of which is that they require a lot of connections. Many connections not only mean more work, it means more resistance and lower efficiency. It also means that these arrays are more difficult to troubleshoot if something goes wrong. Designs I've seen don't pack as many solar cells in as tightly as in crystalline PV modules. The net result of this is that you will need more roof space to produce the electricity required by your home or business.

Solarscapes

Another BIPV option designed to integrate solar electricity into buildings is *solarscaping.* In this approach, PV modules are incorporated into the roofs of shade structures for decks, patios, or even hot tubs. Solar shade structures are typically made from wood or steel and are designed and built so they blend in with the existing architecture and landscape. The PV modules generate electricity while providing shade (Figure 8.29).

Many installers use bifacial PV modules like those made by Sanyo and a few other manufacturers. They harvest solar energy from the front sides from sunlight striking from above and the back sides from sunlight

Figure 8.29. *Solarscaping. PV modules can be used to create roofs of shade structures, a technique known as solarscaping. This photograph shows a portion of a carport made from PV modules by Lumos Solar.* Credit: Lumos Solar.

reflecting either off the ground or off light-colored materials. The solar cells in bifacial modules are encased in glass, both front and back, which allows some light to filter through the array, providing some light beneath.

Pros and Cons of Solarscaping

Like other types of BIPV, solarscaping helps minimize aesthetic concerns of conventionally mounted PV systems. It therefore helps more people find a suitable way to incorporate PVs into their lives. Solarscaping allows placement of arrays in the sunniest locations on sun-challenged properties.

These structures can also be built off site, then very quickly assembled onsite. In addition, they offer ease of access to the array. And, by avoiding installation on the roof, you won't have to worry about removing the array when the time comes to replace roof shingles. Like pole-mounted arrays, this approach results in a cooler and more energy-efficient array.

Conclusion: Getting the Most from Your PV System

If you are interested in installing a solar electric system, it should be clear by now that you have many options. When considering options to mount an array, remember that one of your main goals is to produce as much electricity as possible. Here's a list of all the things you can do to make this happen:

1. Locate your PV array in the sunniest (shade-free) location on your property, orienting the array to true south, setting modules at the optimum tilt angle or installing on an adjustable rack or tracker. Ensure access to the Sun from 8 a.m. to 4 or 5 p.m.!
2. Install a Maximum Power Point Tracking controller for battery-based systems or inverter for grid-connected systems to ensure maximum array output.
3. Keep modules as cool as possible by mounting them on a ground-mounted rack or pole or an elevated roof rack. In hot climates, install high-temperature modules, that is, modules that perform better under higher temperatures.
4. Install high-efficiency modules if roof space or rack space is limited.
5. When installing multiple rows, be sure to provide a sufficient amount of distance between the rows to avoid inter-row shading.

6. Select modules with the lowest rated power tolerance. Power tolerance is expressed as a percentage, which indicates the percent by which a PV module will overperform (produce more power) or underperform (produce less power) the nameplate rating. Look for modules with a small negative or a positive-only power tolerance.
7. Install an efficient inverter. Use the weighted efficiency as your guide rather than maximum efficiency. Weighted efficiency is a measure of the efficiency under typical operating conditions.
8. Keep your inverter cool. If installed outside, be sure the inverter is shaded at all times. If installed inside, select a cool location. Be sure air can circulate around the inverter.
9. Decrease line losses by installing a high-voltage array. Reduce the length of wire runs whenever possible. Use larger conductors than required by Code to reduce resistance losses. Be sure all connections are tight to reduce resistance and conduction losses.
10. Keep your modules clean and snow free. If you live in a dusty environment with very little rain, periodically wash your modules. Remove snow from modules in the winter, but brush it off gently. Don't use any sharp tools to remove snow. Mount the array so it is easily accessible for cleaning and/or snow removal.
11. For battery-based systems, maintain batteries properly. Periodically equalize batteries and fill flooded lead-acid batteries with distilled water. Recharge promptly after deep discharges, and install batteries in a warm location so they stay at 75°F to 80°F, if possible.

CHAPTER 9

Final Considerations

Permits, Covenants, Utility Interconnection, and Buying a System

To those who are enamored by the idea of generating their own electricity from the Sun, there are few things in the world more exciting than turning a PV system on for the first time and watching the meter run backward—meaning you are not only producing all the electricity your home needs, you are producing a surplus that you can use later. I've installed over two dozen PV systems for myself and clients and still get a thrill when we "fire up" a PV system. It's an even greater thrill when you receive your first utility bill and discover the only fee you have to pay is a monthly service charge. It's even more exciting when you find you actually produced a surplus!

Although you may encounter a few obstacles along the way, in most instances, the path from conception to a completed installation is fairly straightforward—and it is getting easier as more and more utilities become familiar with solar electric systems.

To help you on your journey, I've included a list of steps you or your installer must take to install a PV system—in the order in which they must be completed. As you will see, we've already discussed steps 1 and 3 at length. In this chapter, we'll explore the remaining steps. I will begin by discussing permits, then look at restrictive covenants imposed by some homeowners and neighborhood associations. I will then describe interconnection agreements required for grid-connected PV systems and insurance. I'll end with some advice on buying and installing a PV system.

Permitting a PV System

After you or your installer have designed a PV system to meet your needs, you'll need to contact your local building department—managed by the

Steps to Implement a PV Energy Project

1. Determine your home or business' electrical consumption and consider making efficiency improvements to reduce the PV system size and cost.
2. Assess the solar resource.
3. Size and design the system.
4. Check homeowner association regulations; file necessary paperwork for permission to install the system, if required.
5. Apply for special incentives that may be available from the utility or your state or local government.
6. Check on building permit requirements; file a permit application.
7. Check on insurance coverage.
8. Contact the local utility and negotiate utility interconnection agreement (for grid-connected systems).
9. Obtain permit.
10. Order modules, rack, and balance of system.
11. Install system.
12. Commission — require installer to verify performance of the system.
13. Sign interconnection agreement.

city, town, or county—to determine if they require a permit. In most cases, an electrical permit will be required. If, however, you live in a less-regulated state like Missouri, you may find that only a handful of counties require building or electric permits, often only those containing major cities. To find out where your building department is located, ask a local builder or an electrician or call local government. Once you locate the permitting agency, ask for a permit application. If you are installing your own system, they may guide you through the process. If you hire a professional, he or she will secure a permit so you don't have to worry about this detail.

Building departments are known as *authorities having jurisdiction,* or AHJs for short. They control all aspects of construction in cities, towns, or counties. They are granted the authority to administer, interpret, and enforce building codes.

Building codes are a detailed set of regulations that stipulate how buildings are constructed, modified, and repaired. Building codes set the rules by which the various trades like electricians and framers must operate. They stipulate equipment and materials that can be used and how they must be installed. For example, they stipulate the way PV systems have to be wired and the safety measures that must be incorporated. Local trade licensing codes (separate from building codes) may also stipulate who can perform certain tasks, for example, some jurisdictions stipulate that certain aspects of a solar installation must be performed by licensed electricians. Building departments may also stipulate setbacks—that is, how close a PV system may be installed to a public right of way or a neighbor's property line.

Building departments may require an electrical permit for a PV installation and, for building-mounted systems, a letter from a structural engineer certifying that your roof is strong enough to support the PV array. They may also require a structural permit for ground-mounted racks.

Building departments conduct inspections at certain stages during installation. For a roof-mounted PV system, only one inspection, the final inspection, is typically required. For ground-mounted PV arrays, the building department may want to inspect the trench in which you run your wire from the array and the holes you dig for the concrete piers and pads that support the steel rack. They also typically inspect the wiring after the system is installed.

When a project passes all required inspections, the local AHJ will sign off on it. If you are connecting to the grid, you can then notify the utility and request their inspection. When that's completed, you can officially connect to the utility system. (More on this shortly.) For those who are installing off-grid systems, don't think you are free from permitting. An electrical permit will be required for off-grid systems if you live in a city, town, or county that has adopted the National Electrical Code.

Although there's a lot of grumbling about building codes and permits among builders and installers, codes serve an important purpose. They ensure that all building projects, including the installation of PV systems, are safe for the immediate occupants of a home or business—as well as for future residents of a home or employees of a business. The National Electrical Code, for example, protects against potential shock and fires caused by electricity. If you install a PV system that is not in accordance with the Code and it causes a fire that destroys your house, your insurance company may deny your claim. The NEC provisions that apply to PV systems also help ensure the safety of electricians who may come to work on your system and utility company employees (notably, line workers) in the case of utility-connected systems.

Although the requirements of local building codes can be daunting to the uninitiated, professional installers should be intimately familiar with them. If you're going to install your own system, you must learn the Code that applies to your installation and follow it.

Securing a Permit

To start the process, you or your installer submit a permit application if required in your area. Permits for PV systems are pretty simple. They encompass electrical wiring and all the electronic components you'll be installing, such as inverters, charge controllers, and safety disconnects. You may need to include a description of how the array is mounted, so the permitting authority can be sure that your roof can support a roof-mounted array and that it will be securely attached so it won't blow away in the wind. (More on this shortly.) Your building department will outline the procedure, indicate the cost, and provide a form to fill out. They will also indicate all the supporting material you'll need to provide in your application.

Applications may require a site map, drawn to scale, that indicates property lines, streets, and the proposed location of the array (Figure 9.1). If required, this drawing should also indicate the location of other components of the system, for example, the inverter, disconnects, and electric meter (in grid-connected systems). A professional installer will prepare a site map for you. It's part of his or her service.

Although building permit applications may not require a site map, it's a good idea to include one. All building departments will require a one-line drawing (aka a single-line drawing). As shown in Figure 9.2, a one-line drawing shows all the components of the system, including the PV array. It also lists the specs of each component and illustrates wire runs, indicating the size of wire required to comply with the NEC. One-line drawings also show how the system will be connected to the utility grid. In some jurisdictions, AHJs may require more complex three-line electrical diagram. These drawings show all the wires, including ground wires.

You may also need to provide a description and/or drawings of the rack when applying for a permit. If you're mounting an array on the ground, you may need to provide drawings of the foundation and the rack. If you

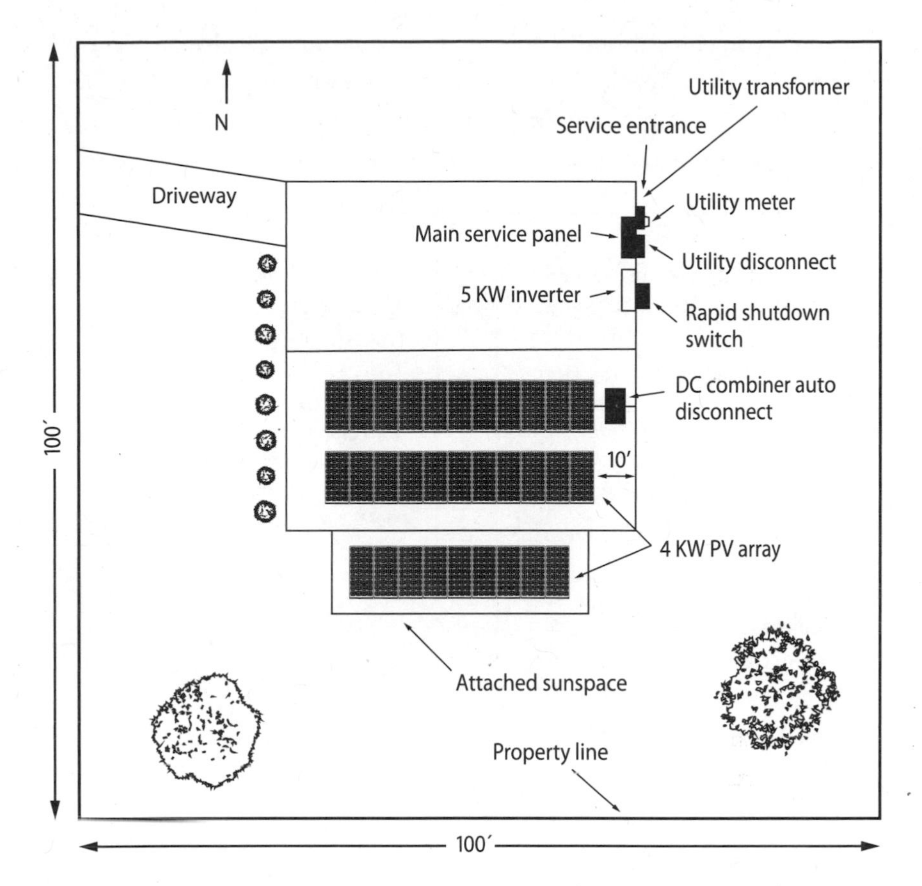

Figure 9.1. *Sample Site Map. A site map like this one shows the location of the array and important equipment. It may be required by your building department.* Credit: Forrest Chiras.

are mounting an array on a roof, some AHJs require information on the age of the roof, type of shingles, the pitch, and the size and spacing of the rafters. You may also need to provide information on the weight of the array and the method of securing it to the roof. You may need to provide details on waterproofing the attachment as well, although that's unlikely. Because building departments may not want to determine whether your roof can support the weight of an array, they may ask you to hire a licensed structural engineer to review and approve ("stamp") your plans. An engineer's stamp ensures that the roof can support the array, even if your roof is blanketed in snow. It also ensures that it will secure the array properly based on local wind conditions.

Even if your local building department does not require an electrical permit, which is rare, don't think you are out of the woods. Your local utility *will* ask for a one-line drawing and spec sheets for modules and inverters. In addition, because many utilities often don't have the personnel

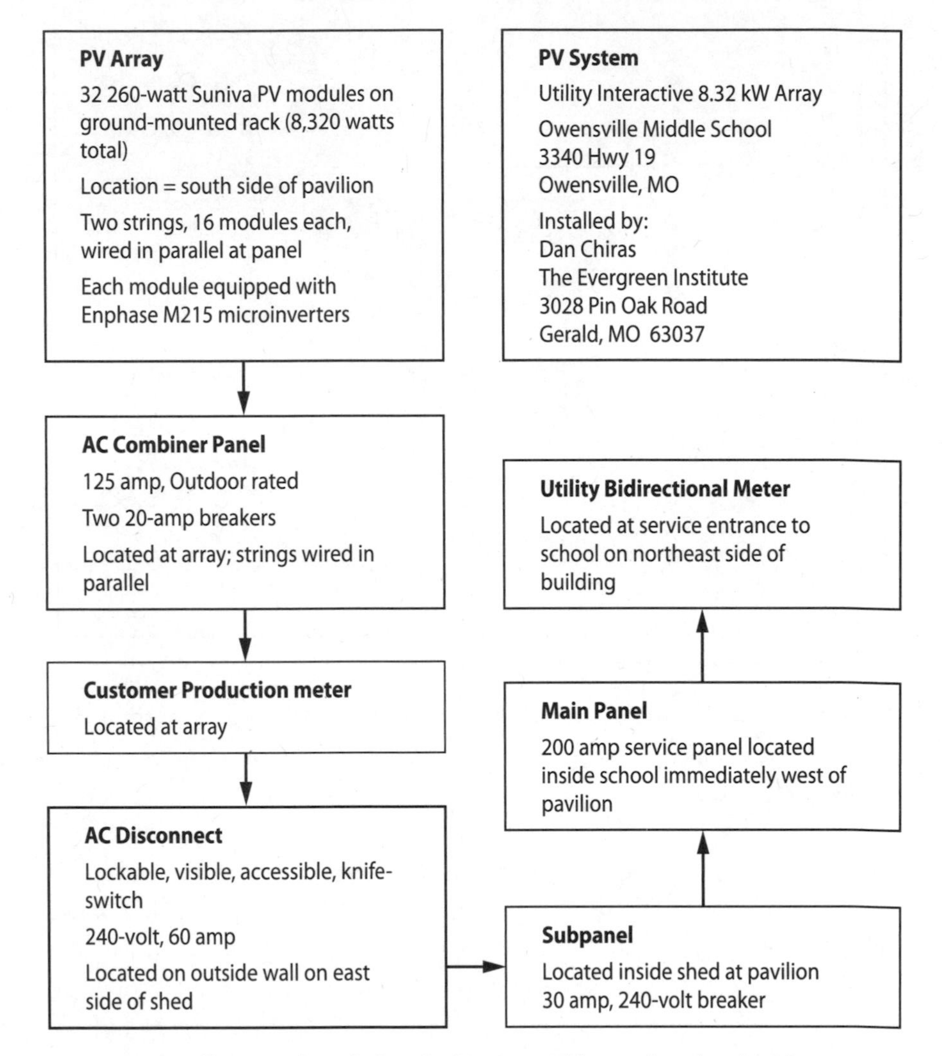

Figure 9.2. *One-line Drawing of Electrical System. This one-line drawing shows the layout of a PV system, including all the equipment that will be installed and pertinent details about that equipment. Wire sizes are also included. Building departments always require these drawings, and utilities may request one as well. Be sure to check your AHJ and utility first to see the level of detail they ask for. In my experience, the more detail, the better.* Credit: Dan Chiras.

trained to assess PV systems, they typically ask the installer to submit a pre-construction certification—a statement from a licensed electrician or engineer signifying that the system was designed in compliance with the National Electrical Code.

After your application is submitted, a plan examiner in the building department will review the application and accompanying materials, usually within a few days or weeks, although the process can take several months in departments that are inundated with permits. If everything is in order, you'll receive a permit, an official approval for you to commence construction. (Never order equipment or start work until you've received your permit!)

If your permit application doesn't meet their approval, the plan inspector will mark up the schematics—it's called *redlining*—or provide written notice indicating the problem or problems and required modifications. Once the problems have been resolved, you must resubmit the permit application.

Fees for permits for residential PV systems usually run from $50 to $1,500, depending on the jurisdiction. Some jurisdictions charge a flat fee; others charge a fee based on the size of the system. Permits must be posted on the site, usually in a window. Expect an electrical inspector to visit your site at the end of an installation to check wiring and warning labels. (Warning labels are required on disconnects and other components of PV systems.) The inspector may also check wire sizes, overcurrent protection, and grounding.

All equipment must be readily accessible to inspectors and must be installed according to Code with proper clearances (to ensure room to work on the equipment). This is known as *working space.* You'll also need to be sure that no other wiring, plumbing, or ductwork is within a certain distance of installed equipment. You or the installer have to call the AHJ to schedule all inspections, usually 24 to 48 hours in advance.

If you fail an inspection, you'll need to fix the problem and arrange a follow-up inspection. You may have to pay for follow-up inspections, too, although the cost of follow-up inspections is usually included in a one-time permit fee.

When applying for a permit yourself, be sure you follow all guidelines or hire a professional installer who does! Be courteous and respectful in all your dealings with the building department. If you are feeling irritated at having to file and pay a fee, leave your angst at home. The same goes for dealing with inspectors. Inspectors have a tough job. They deal with difficult electricians and builders day in and day out and often show up to a site in a sour mood. Be polite and respectful, even friendly—even if they fail to pass your installation. Ask for explanations on ways to correct mistakes. They're usually happy to help out. When applying for permits or dealing with inspectors, however, expect to be treated by building department officials in a reasonable and timely manner.

While we are on the subject, avoid the urge to bypass the law—to install a system without a permit. Do not, under any circumstances, try to sidestep this requirement. The consequences are too great. Municipal governments have the authority to force homeowners to remove unpermitted structures, even entire homes. Be sure to sign an interconnection agreement with the local utility, too. With all the sophisticated monitoring in place, the utility will know if you are backfeeding electricity onto the grid. If you live in a county that does not require permits, be sure to install your system to Code. Lack of enforcement is no excuse for shoddy, unsafe installations.

Permitting a PV system may take several months, so submit your application well in advance of the date you'd like to install the system, *and don't buy a PV system until your permit has been granted.*

Covenants and Neighborhood Concerns

Installing solar electric systems in suburban neighborhoods can sometimes be challenging. That's because most of them are "governed" by a set of rules. These rules are called *restrictive covenants.* They dictate what can and can't be done to your home or property. Even with a permit from the local building department, restrictive covenants can block a solar installation (although this is pretty rare these days).

Covenants are established when a subdivision is created. In legal terms, covenants "run with the land," meaning they apply to the original owner and as well as subsequent owners of the property. Some covenants give the neighborhood association the right to create and enforce additional rules and architectural standards.

Restrictive covenants and neighborhood association rules may address many aspects of our homes—from the color of paint we can use to the installation of a privacy fence. Some neighborhood associations even ban clotheslines. Some expressly prohibit renewable energy systems, such as solar electric systems.

Restrictive covenants are legally enforceable, and the courts have consistently upheld their legality. To see if you will be prohibited from installing a PV system, you need to review the restrictive covenants *and* neighborhood association rules, if any. Contact your neighborhood association to see what rules apply.

Even if your subdivision has restrictive covenants or architectural standards that allow solar installations, you will need to submit an application to the homeowner's association. The application is often a letter with a drawing indicating where the solar array will go. Precedence can help. In other words, if someone else has installed a similar system, even without permission, it's easier to obtain approval.

Even in the absence of restrictive covenants, you might want to talk to your neighbors about your plans. Many people feel very strongly about protecting the aesthetics of their neighborhood—and the resale value of their home. So, it's a good idea to discuss your plans with your closest neighbors, long before you lay your money down.

Don't expect your neighbors to be as enthusiastic about a PV system as you are. Not everyone is enamored of solar. Some people have a knee-jerk reaction against it. To help win over reluctant neighbors, you may want to show your neighbors pictures of the type of PV system you're going to install. If you are good at Photoshop, take a photo of your home from their house and place a picture of the PV system on your roof or on your property so they can see what it will look like. If you are thinking about installing a grid-connected system, let them know that you'll be supplying part of their energy, too. Giving neighbors advance notice, answering questions, and being responsive to their concerns is the best way to ensure peace in your neighborhood.

If you're building a new home in a historic district or adding a PV system to an existing home in a historic district, you may have a battle on your hands. A building-integrated PV system may be the only possible option, and even that may not meet approval. However, if you can locate the array where it is not visible from the street, your chances for approval should increase.

Covenants can be a pain. When I published the first edition of this book, two renewable energy-friendly states—Colorado and Wisconsin—prohibited homeowners associations from preventing homeowners from installing solar and other renewable energy systems. Today, about half the states in the nation have similar laws. If you're in one of those states, count yourself lucky. If not, you might want to work to pass such legislation in your state.

Connecting to the Grid: Working with Your Local Utility

When installing a grid-connected system, you'll need to contact your local utility company early on—*well before* you purchase your equipment—to secure a copy of their interconnection agreement. It's a good idea to call your utility and speak to the person in charge of grid interconnection. Be courteous. These people can be great allies!

I've dealt with a lot of utility companies. They are mainly concerned that PV systems will backfeed electricity onto the grid when it's down. Because UL 1741-listed grid-tied inverters shut down automatically when the grid goes down, all the utility generally needs is documentation showing that your inverter is UL 1741 listed. (Inverter manufacturers put this listing on their spec sheets; some manufacturers issue certificates of compliance with UL 1741 that you can download from their websites.)

As noted in Chapter 5, most utilities require the installation of visible, lockable AC disconnects—that is, manually operated switches that allow utility workers to disconnect your system from outside your home in case the grid goes down and they need to work on the electrical lines. As you know by now, grid-connected inverters *automatically* disconnect from the grid when they sense a drop in line voltage or a change in frequency, so a utility disconnect is an unnecessary addition. Regardless, it will still be required by most utilities. There's no way around this requirement.

Utility interconnection agreements also require information on the size of the system and the inverter make and model. In addition, interconnection agreements spell out the company's net metering policy, which is dictated, in most cases, by state law. This agreement will explain how the utility handles net excess generation—that is, whether they have annual or monthly reconciliation and how much they reimburse customers for net excess generation.

Interconnection agreements may suggest the purchase of a liability insurance policy. You will probably be shocked when you see that many companies suggest that homeowners secure $1,000,000 liability insurance policies. It's meant to protect the utility from damage your PV system might cause to their workers, their system, or your neighbors—all of which are deemed highly unlikely by experienced solar professionals. Few people secure policies of this size, as you shall soon see.

Interconnection agreements also include a pre-construction document that stipulates that the system will be installed according to Code. They also include a post-construction agreement, which is a form submitted to them *after* the system has been installed. It must be signed by an engineer or licensed electrician who attests that the PV system was installed according to Code. Although both these documents need to be signed by an electrician or engineer, finding one who understands NEC requirements for PV systems can be difficult, especially in rural areas and small towns.

While all this may sound daunting, remember: if you are hiring a professional installer, he or she will take care of the paperwork for you. If you are going it alone, you'll be happy to know that the forms are relatively easy to fill out. You will, however, have to hire an electrician or engineer to sign the pre- and post-construction documents.

Some utilities charge an interconnection fee. This fee may be based on the size of the system, or it may be fixed—with one price covering all systems. Fees typically range from $20 to $800. They cover costs borne by the utility, such as the cost of inspecting a PV system after it is installed, and, more likely, the cost of installing a bidirectional meter. Be sure to ask about interconnection fees upfront, and be sure the contract is clear about any additional fees that could be charged to you over the course of the contract.

Making the Connection

As noted earlier, once a grid-connected PV system has passed final inspections by the local AHJ, you or your installer arrange for the utility to do their inspection. They will inspect the system, change out the meter, if necessary, and sign the interconnection agreement, granting you the right to connect to the grid. *Only then* can you flip the switch and start generating electricity. In some cases, utilities may test a PV system to be sure the automatic disconnect functions as it is supposed to.

Whatever you do, don't switch a PV system on *before* the utility has approved your system for interconnection. While it may take a week or ten days to get the utility to show up at your home, and it is tempting to start producing electricity before they get there, please resist the nearly overwhelming temptation. Although an installer can connect for brief periods to test a system to be sure everything is working, he or she must disconnect until the utility has inspected the system and signed off on it.

So, once the installer tests the system and leaves the premises, resist the temptation to flip the switch to fire it up.

Insurance Requirements

Insuring a solar electric system is a good idea. This is a large investment, and you'll want to protect it. Two types of insurance are advised: (1) property damage, which protects against damage *to* the PV system, for example, a tree falling on it, or vandalism; and (2) liability insurance, which protects you against damage *caused* by your system, for example, an electrical shock someone might receive. Both are provided by homeowner's or property owner's insurance policies.

Insuring Against Property Damage

For homeowners, the most cost-effective way to insure a new PV system against damage is under an existing homeowner's insurance policy. Businesses can cover a PV system under their property insurance, too.

When installing a PV system on a home or outbuilding—or even on a pole-mount—contact your insurance company to determine if your current coverage is sufficient. If not, you'll need to boost the coverage enough to cover replacement, including materials and labor. If you are installing a PV array on a pole, be sure to let them know that the PV system should be insured as an "appurtenant structure" on your current homeowner's policy. This is a term used by the insurance industry to refer to any uninhabited structure on your property not physically attached to your home. Examples include unattached garages, barns, sheds, and satellite dishes.

Insuring a PV system is relatively inexpensive. Systems mounted on homes are generally covered, but check to be sure. Home insurance often covers appurtenant structures like sheds, although you may be required to purchase additional coverage for an additional premium.

Liability Coverage

Liability insurance protects against possible damage to others caused by a PV system. Although I have never heard of any such damage occurring, liability insurance is worth having.

Liability insurance covers possible claims from damage to a neighbor's electronic devices from a grid-connected PV system. (For example, if your system sends power onto the line that somehow, magically, damages electronic equipment in a neighbor's home.) It also covers personal injury or death of employees due to electrical shock from a system when working on a utility line during a power outage. Even though the likelihood of this is nil because of the automatic disconnect feature built into inverters, utilities may recommend this coverage. However, as noted earlier, they typically suggest liability coverage way in excess of what's reasonable.

Liability coverage is part of a homeowner's or business owner's policies. In most places, liability coverage for homes provides protection from $100,000 to $300,000.

Buying a PV System

If a PV system seems like a good idea, I strongly recommend hiring a competent, bonded, trustworthy, and experienced professional installer. A local installer will supply all of the equipment, file for permits and interconnection agreements, and install the system. Your installer will test the system to be sure it is operating correctly (don't pay in full until you are satisfied). Be sure to obtain written documentation of equipment warranties and any guarantee the installer makes regarding the system.

Consider hiring an installer who's bonded, especially if he or she is new to the business. A bond is sum of money set aside by contractors and held by a third party, known as a *surety company.* The bond provides financial recourse to homeowners and business owners should a contractor fail to meet his contractual obligations.

When contacting companies, ask if they are bonded and for how much. Check it out. Be sure the installer has worker's compensation insurance to protect his or her employees when they are working on your site. Also ask for references, and be sure to call them. Visit installations, if possible. Talk to homeowners about their experience and level of satisfaction with the installer, his crew, and the array. You may also want to check with the local office of the Better Business Bureau to check out an installer's rating. And, of course, get everything in writing. Sign a contract.

Whatever you do, don't pay for the entire installation up front. I ask for one-third upon signing a contract, so long as I can get to the installation within two to three weeks. That is the amount of time it takes to order and have equipment shipped to me. Getting a permit is pretty easy these days, and installers will take care of that detail. It's pretty safe to provide a deposit after an installer has given you a bid.

I ask for another third when I show up to begin the installation, and the last third when the building department and utility okay the system and it is up and running. See the sidebar, "Questions to Ask Potential Installers," for a list of questions you should ask when shopping for a competent installer.

Questions to Ask Potential Installers

1. How long have you been in the business? (The longer, the better.)
2. How many systems have you installed? How many systems like mine have you installed? (The more systems, the better.)
3. How will you size my system?
4. Do you provide recommendations to make my home more energy efficient first? (As stressed in the text, energy-efficiency measures reduce system size and can save you a fortune.)
5. Do you carry liability and worker's compensation insurance? Can I have the policy numbers and name of the insurance agents?
6. Are you bonded?
7. What additional training have you undergone? When? Are you NABCEP certified? (Manufacturers often offer training on new equipment to keep installers up to date; and NABCEP is a national certifying board that requires installers to pass a rigorous test and have a certain amount of experience.)
8. Will employees be working on the system? What training have they received? Will you be working with the crew or overseeing their work? If you're overseeing the work, how often will you check up on them? If I have problems with any of your workers, will you respond immediately? (Be sure that the owner of the company will be actively involved in your system or that he or she sends an experienced crew to your site.)
9. Are you a licensed electrician or will a licensed electrician be working on the crew? (State regulations on who can install a PV system vary. A licensed electrician may not be required, except to pull the permit, supervise the project, and make the final connection to your electrical panel.)
10. Will you pull and pay for the permits?

11. Will you secure the interconnection agreement?
12. What brand modules and inverters will you use? Do you install all UL-listed components? (To meet Code, all components must be either UL-listed or listed by some other similar organization.)
13. What are the warranties on each of these components? Can I have copies of the warranties?
14. What is the payment schedule?
15. Do you guarantee your work? For how long? What does your guarantee cover? How quickly will you respond if troubles emerge? (You want an installer who guarantees the installation for a reasonable time and who will promptly fix any problems that arise.)
16. Can I have a list of your last five projects with contact information? (Be sure to call references and talk with homeowners to see how well the installer performed and how easy he or she was to work with.)
17. What's a realistic schedule? When can you obtain the equipment? When can you start work? How long will the whole project take?

As a rule, I don't recommend installing a PV system yourself. To the inexperienced, Code requirements can be daunting. You really need to know your stuff just to secure a permit, let alone install the system safely. And, there are a lot of ways to mess up, especially with system grounding. Self-installs are appropriate for those who know a fair amount about electrical wiring and have lots of experience. You can gain some of this experience by attending several hands-on PV installation workshops.

Installing a PV system can also be risky. Working on electrical wiring is fraught with potential shock hazards. Connecting to the electrical grid is a job best left to professionals, because working inside a main service panel can be extremely dangerous. Remember, a PV system is a huge investment and so are you, and you don't want to mess either up.

Another option is to work with an educational center like the Midwest Renewable Energy Association. They're often looking for homeowners who are willing to sponsor a hands-on installation workshop. Workshop registrants pay a fee to cover the cost of the instructor, the costs of advertising and setting up a workshop, and perhaps renting a portable toilet.

The homeowner pays the costs of materials, and the workshop attendees provide free labor.

Workshop installations can be fun and satisfying. You get a PV system installed on your home or business while providing a valuable opportunity for others to learn. You do, however, have to be comfortable with a half dozen to a dozen or more people working on your site for four or five days. If the workshop leader is competent and checks all of the attendees' work, you'll get a PV system installed right. Be sure to contact the nonprofit organization well in advance and be prepared to help organize and publicize the workshop. Also, be sure your insurance will cover volunteer workers on your site.

Parting Thoughts

When you started reading this book, you no doubt were already interested in PV systems. Perhaps you just wanted to determine if PV would be suitable for your home. Perhaps you were sure you wanted to install a PV system but didn't know how to proceed. I hope that you now have a clearer understanding of what is involved.

You know a great deal about solar energy, electricity, PV modules, types of PV systems, inverters, batteries, charge controllers, racks and mounting, and permits. You've gained good theoretical as well as practical information that puts you in a good position to move forward.

My work has ended, but yours is just beginning. I wish you the best of luck!

Index

Page numbers in *italics* indicate figures.